Grape vs. Grain

Why is wine considered more sophisticated even though the production of beer is much more technologically complex? Why is wine touted for its health benefits when beer has more nutritive value? Why does wine conjure up images of staid dinner parties while beer denotes screaming young partiers?

Charles Bamforth explores several paradoxes involving these beverages, paying special attention to the culture surrounding each. He argues that beer can be just as grown-up and worldly as wine and be part of a healthy, mature lifestyle.

Both beer and wine have histories spanning thousands of years. This is the first book to compare them from the perspectives of history, technology, nature of the market for each, quality attributes, types and styles, and the effect that they have on human health and nutrition.

Charles Bamforth is Chair of the Department of Food Science and Technology and Anheuser-Busch Endowed Professor of Malting and Brewing Sciences at the University of California, Davis. He is Editor-in-Chief of the *Journal of the American Society of Brewing Chemists* and a member of the editorial boards of the *Master Brewers Association of the Americas Technical Quarterly*, the *Journal of the Science of Food and Agriculture*, and the *Journal of the Institute of Brewing*. Bamforth is one of the top two or three brewing scientists of his generation.

Grape vs. Grain

A Historical, Technological, and Social Comparison of Wine and Beer

Charles Bamforth
University of California, Davis

CAMBRIDGE
UNIVERSITY PRESS

CAMBRIDGE UNIVERSITY PRESS
Cambridge, New York, Melbourne, Madrid, Cape Town, Singapore, São Paulo, Delhi

Cambridge University Press
32 Avenue of the Americas, New York, NY 10013-2473, USA

www.cambridge.org
Information on this title: www.cambridge.org/9780521849371

© Charles Bamforth 2008

First published 2008

Printed in the United States of America

A catalog record for this publication is available from the British Library.

Library of Congress Cataloging in Publication Data
Bamforth, Charles W., 1952–
Grape vs. grain : a historical, technological, and social comparison of wine
and beer / Charles Bamforth.
p. cm.
Includes bibliographical references and index.
ISBN 978-0-521-84937-1 (hardback)
1. Brewing. 2. Wine and wine making. I. Title.
TP570.B1827 2008
641.2′1–dc22   2007029237

ISBN 978-0-521-84937-1 hardback

For Charlie's Angels

CONTENTS

PREFACE

I FLEW TO HEATHROW FROM INDIA, VIA FRANKFURT. THE FOUR-hour holdover in the German airport had not remotely bothered me. I hate tight connections, and, besides, I was able to indulge in some sausages and weissbier while peaceably reading my newspaper, a faint buzz of conversation surrounding me.

Later the same day, I found myself for the first time in several years in central London. Strolling toward Hyde Park Corner in the dusk of early evening, it occurred to me that the traffic heading toward the West End was much heavier than I recalled from when I was a more regular visitor and living just a short train ride away. As I walked, there was suddenly the most stupendous whooping, and I turned to see two girls, probably late teens, hanging (in every sense of the word) out of the windows of a stretch limo and gyrating maniacally.

I thought little of it – surely an aberration – and continued my stroll, eventually pitching up at The Crown on Brewer Street, close to Piccadilly Circus. It was a hostelry I knew of old, and, in truth, little within had changed, with the exception of the display on the bar. There was row upon row of taps for dispensing kegged beer, but just a solitary handle for pumping traditional English ale from the cask. I had a pint of the latter, a worthy drop of Charles Wells Bombardier.

Half an hour later, I took a table at an Italian restaurant on Wardour Street and washed some crisp salad and succulent lamb's liver down with successive glasses of the house white and house red, both charming Italian vintages. There wasn't much on offer by way of beer.

Venturing back toward the Underground station, I decided to take in one more pub, this the St. James Tavern on Great Windmill Street (being of *that* age, I needed the loo more than I needed another pint). The bouncer on the door looked at me curiously but said little. I soon realized why. I, an amply-bellied and balding fifty-something, must have looked like a solitary cashew amidst a heap of raisins. The place was heaving. Extremely young people were screaming to be heard above a blast of decibels that must surely have been making their ears bleed. There wasn't a glass in sight; rather, everyone was hugging to their bosoms (no space for arm's length here) bottles of premium lagers or RTDs ("ready to drinks"), also known as Malternatives or FABs ("flavored alcoholic beverages"). I craned my neck to look at the bar, but saw no immediate evidence of beer pumps. Feeling claustrophobic, I made for the restroom. Through the door of the cubicle, I saw that the toilet had collapsed in pieces. Almost in panic, I wrestled my way back to the bouncer. "I think I jut doubled the average age in here." He smirked and looked away.

Back on the street, as I gasped my deepest for breath, another stretch limo crawled past amidst the jam of cars, incongruous rickshaws, and people spilling off the pavement. From the limo windows, young girls caterwauled.

As I sat, dumbfounded, on the tube train taking me back towards my hotel, it occurred to me how I had that day witnessed living proof of a thesis that forms the heart of this book. Early twenty-first century London is the embodiment of why alcohol, most especially beer, has achieved such a negative image in the minds of many.

In the space of less than a day, I saw examples of the decline of locally traditional values in a proud industry (the dearth of cask ale in London) and beheld the positioning of wine in a more refined and wholesome environment (the restaurant, which had two pages devoted to the wine selection, and just one line to the solitary bottled beer available). Yet, I had confirmed how beer (the wheat beer in Germany) can also be dealt with respectably as well as being the perfect accompaniment to a meal. I experienced the evolution of the current drinking ritual, which has little if anything to do with the quality criteria that I describe for wine and beer in this book and everything to do with displays of sexual and other forms of aggression, and addiction to partying.

The producers of alcoholic drinks get the blame. In fact, it's the *purveyors* of alcohol that are letting everyone down, both by enticing the young into such displays of abandon and for not emphasizing and marketing the genuine wholesome attributes of the alcoholic beverages that have become secondary to the real fixation. Moderation is not a word in the lexicon of these people.

This book is about beer and wine. It speaks of the worthiness of each as part of a respectable, respectful, and restrained lifestyle. Above all, though, based on the unfortunate belief of many that beer is a "bottom feeder" in the world of alcohol, with wine bobbing on waves of respectability, I seek to compare these two beverages on the basis of their history, technology, scientific and artistic appreciation, and impact on the body. To that extent, and reflecting my professional specialty, the theme of this book is primarily one of demonstrating how beer is a product of an excellence and sophistication to match wine, and I seek to do this by championing beer while being entirely fair to that other noble beverage.

Grape vs. Grain

1. Beer and Wine

Some Social Commentary

I MUST COME CLEAN BY ADMITTING TO HAVE WORKED IN OR around the brewing industry for nearly thirty years. It will come as no surprise to you, then, that I drink beer. I like beer. I admire brewers. I think they are some of the most skilled, devoted, and ingenious people on the planet. Charming, too.

However, I do not dislike wine, nor the viticulturalists and enologists who bring that amazing product to the market. I drink wine, though I prefer beer. I believe that the brewer has much to learn from the winemaker with regard to re-establishing their product as an integral component of a wholesome and elongated lifestyle. Equally, the winemaker must doff his or her cap to the brewer insofar as technical matters go. There is no question that brewing leads the way in matters technological and scientific. Indeed, throughout the industrial ages, brewing has been a pioneering process that has informed all other fermentation industries, even to the production of pharmaceuticals and the latter-day *biotechnologicals*, with their diversity of high-value products.

In this book, I compare beer and wine. I do not seek to decry wine. Rather, I aim to demonstrate why brewers can hold their heads high in the knowledge that their liquid is every ounce the equal of wine, by any yardstick you choose to nominate. At the same time, I will highlight the frustrations I have that many brewers do so little to truly champion beer for its inherent qualities, but rather seek the popularity low ground with their outrageous (if hilarious) advertising regimes and product innovation strategies that spawn drinks (notably the Malternatives) that are alarmingly variant to the beers that we have enjoyed for generations. Compare, if you will, the imagery associated with beer as opposed to that of wine.

The word "wine" conjures images for me of chateaux on hillsides in leafy France. Or I picture the mission-style façade of Robert Mondavi's vineyard, as well as wineries of other notable names in Napa. I see the ritual of the wine sommelier rejoicing in the ceremony of the bottle presentation, ritualistic decorking, and offering for approval. I applaud row upon row of books on wine at any bookstore, rich volumes held in biblical reverence. I see family picnics with wicker baskets containing canapés and smoked salmon, with ladies in gossamer gowns.

As the reader will deduce from this book, I have no doubt in my own mind that beer could just as readily occupy rarefied scenarios, but much more besides. Frankly, though, a word association game with the word beer will generate for many images of steaming factories in dark mill towns, bottles with torn labels plunked on Formica tables to accompany burgers and fries and perhaps a cigarette. Likely as not, your waiter will not spontaneously give you a glass to drink from and, even when you get one, it is likely not the appropriate receptacle for the beer concerned and will be badly washed, leading to instant killing of the foam or at best some dreadful bladdery bubbles. And that is if you get to see the bubbles, for the waitress is likely to go to extreme lengths to pour real gently down the side of the glass so as to *avoid* bubble formation. When will they learn that they should splash the beer down into the center of the base of the glass, so as to give the carbon dioxide every opportunity to convert to bubbles, and so that a large head of foam will be produced? *Then* they can leave you with the bottle for you to top up your glass to your heart's (and eye's and mouth's) content.

I struggle to find books on beer, and when I do they inevitably turn out to be treatises on how to brew your own in a bucket or a thousand-and-one ways to describe the state of drunkenness. Words like suds that have crept into common parlance are better suited to bowls of washing, not beer. I see gangs of loud immature men, belching, farting, and falling over. I see drinking games and ritualistic bad behavior.

I am not naïve. I know which segment of society drinks the most beer: young males. Not in a million years would I presume to suggest that brewers should turn their backs on this sector. But I trust that they will preach moderation to them. My wish is that all brewers will realize (as some do already) that they too can appeal to those who presently savor their wines and who perhaps do not realize the pedigree and potential of the grain-based beverage.

Travel the world, though, and you will find cultures where beer is very much romanced and regarded as more than an equal for wine; indeed, it is an inherent cultural feature. Nowhere is this truer than Belgium. The diversity of beers is immense: There is not an occasion unsuited to the consumption of beer, with the exception of the communion chalice. Not only that, the beer must be presented with a theater and reverence no less essential than for the pouring of a fine wine.

I recall a visit to the home of a good friend in Antwerp. He invited me to choose a pre-prandial beer from his collection and I dutifully selected something after hearing the seemingly unending list of what he had to offer. The drink did not appear for twenty minutes as he hunted for exactly the correct glass, the one with the appropriate shape and dimensions for the beer, the one with the relevant brand logo in place. I told him that it really didn't matter which glass he provided, but I might as well have told him that it was immaterial whether the subsequent excellent meal was served on paper plates and eaten with plastic knives and forks.

Thus, Belgium falls fair and square in the list of "beer countries." Alongside them we might also list the likes of the Czech Republic (home to comfortably the most formidable drinkers of beer), Germany, Ireland, the United Kingdom and, yes, the United States (Table 1-1). "Wine countries" include France and Italy. Perhaps it will come as a surprise to some, though, that in Spain and Portugal, they consume more liters of beer than they do of wine. Of course, we must factor in the strength of the beverages if we are to consider any country in terms of the amount of alcohol derived from the two sources.

Perusal of Table 1-1, however, reveals the magnitude of the difference between beer and wine for their importance to the American customer. There is a slight decline in beer and increase in wine over the five-year period displayed, but beer is still the premier beverage in a nation that represents the second largest beer market (after China – where there is not much consumption per head, but an awful lot of people). The sheer population of the United States itself means that, despite the low per capita consumption of wine, it is the third largest wine market in the world (after France and Italy).

There are not many countries where beer volumes are growing. China has been a phenomenal story: From 1970 to 2003, the annual production of beer increased from 1.2 to 251 million hectoliters. To

TABLE 1-1. BEER AND WINE IN MAJOR MARKETS

Country	Beer consumed (liters per head)		Wine consumed (liters per head)	
	1998	2003	1998	2003
Australia	95.0	87.3	19.7	20.7
Belgium	99.0	96.6	21.7	24.8
Brazil	50.2	45.9	1.4	2.0
Canada	67.0	68.4	8.9	11.0
China	15.6	19.4	0.9	0.9
Czech Republic	160.8	161.0	15.4	16.9
Denmark	107.7	96.2	29.1	30.1
France	38.6	35.5	58.1	49.0
Germany	127.4	117.5	22.8	23.6
Ireland	124.2	118.0	8.8	13.3
Italy	26.9	30.1	52.0	50.5
Japan	57.2	50.9	3.3	2.2
Mexico	49.0	51.7	–	–
Netherlands	84.3	78.7	18.4	19.6
Portugal	65.3	60.0	58.0	50.0
Russia	22.5	51.4	6.0	8.2
Spain	66.4	78.3	35.0	28.2
United Kingdom	100.6	101.3	15.7	20.0
United States	83.7	81.6	7.3	8.0

Data courtesy of the British Beer and Pub Association.

put that in context, consider that the U.S. production went from 158 to 230.8 million hectoliters. Per capita consumption in China, while increasing, remains low, for disposal income is still limited. Countries where beer consumption per head is growing include Russia (and other countries from what was once the USSR – vodka is losing its compulsion), Spain, and South Korea. Several countries are showing steady growth in wine consumption per head, as a glance at Table 1-1 shows. The decline in wine consumption is notable for being located in the great wine countries of France, Spain, Italy, and Portugal. Beer in decline and wine on the increase in Germany, beer on the rise and wine in decline in Spain: a case of seeing how the other half lives?

The factors impacting the overall consumption of alcoholic beverages in any community are diverse and complex. They include lifestyle and consumer demographics, notably age, disposable income, and, of course, image. In many societies, the populace is becoming better educated, wealthier, and choosier. They expect choice and diversity. They make decisions based on perceived quality. Balance judgments are made and, in this, surely the wine industry in countries such as the United States has stolen the moral high ground. Where there is growth in the beer market in this country, it reflects the consumer's interest in matters of health and well-being, ergo the march of light beer, or an interest in products different from the mainstream lighter flavored lager-style beers that still comprise the bedrock of the industry. Thus, the emergence of the craft brewing industry spoke to the interest of consumers in regional, fully flavored (I would say in many instances grossly *over*-hopped) products. The search for new types of beer also fueled the demand for imported products that capture the imagination and assume a provenance of traditional British, Germanic, Irish, or Dutch brewing excellence. The fact remains that the majority of beers imported into the United States display a profoundly aged character, and assume an aroma described by the beer taste expert as wet paper or cardboard. To the cognoscenti, this is as reprehensible as corked taint in wine. But the consumer still buys these beers, purchasing a bit of Burton, a dash of Dublin, or a schloss of Stuttgart.

I am quite convinced that customers purchasing imports and regional brews have in their mind's eye an image of a rustic brewery handcrafting beer in time-honored vessels according to ancient recipes. In reality, many of the imports are from huge modern breweries employing the latest technologies and practices. Some of the micro-brews in the United States emerge from dubious equipment, badly configured and begging for investment. As such, the quality of some of these beers as judged from within the expert brewing community is, frankly, deplorable. By contrast, the domestic brands produced in huge volumes and flashed across our television screens between baseball innings and at basketball time-outs are outstanding for their consistency, cleanness, and purity. Yet it is the micro-brews that have captured the consumer's imagination by touting a perceived sophistication that belongs in the same class as that engendered by wines.

There are major global beer brands – not least Guinness – that have assumed an aura of romance. I recall some while ago receiving an e-mail from a woman who asked whether it was true that the difference between Guinness brewed in London and that in the mother brewery of Dublin was that they marinate a dead cow in the brew from Ireland. They don't.

The major breweries of the United States and elsewhere in the world are places of sophistication and excellence. They are hygienic, airy, busy, and highly productive. They are working 24 hours a day, 365 days in the year, striving towards products of consistent excellence. They may be packaging bottles at rates exceeding 1,200 bottles every minute or cans more than twice as quickly as that. And most often, they are located in less than beautiful surroundings, historically matched to centers of high urban population. To the consumer, these are factories, but you will never hear a brewer using that word. They are breweries, sophisticated, often highly automated, but always using time-honored brewing techniques. The only differences from a pub-based brewery are that the brewer spends more time looking at a computer screen, the facility is almost always rather cleaner, and the product is invariably more consistently excellent.

Occasionally, one encounters a brewery of genuine esthetic appeal. For example, you should head to the Sierra Nevada brewery in Chico, California, to see a facility that is the equal of any winery for style and sophistication, while at the same time having a technological and environmental conscience unmatched by the vast majority of winemakers.

There are some wineries that aspire to the sophistication of the mega breweries. And in just the same way that the major brewers are too often decried for producing somehow inferior products (when the reality is that their beers are actually vastly superior in terms of consistency of quality), the big "factory" wineries are pilloried as churning out down-market hooch. Nothing could be further from the truth, yet it is even more the case for wine than for beer that the customer associates quality and excellence with the products of smaller rural wineries of charm and elegance. It is almost as if the quality of the wine derives more from the art of the architect rather than the wit of the winemaker.

Winemakers speak of vintage. The plethora of wine literature grinds the dust fine on nuances of year, varietal, and winery. Brewers don't: A brand is a brand, and the expectation is that the beer should taste as expected every time, year on year. No need, then, for the bartender to offer the customer a sniff of the beer first to garner agreement that the rest is fit for consumption. No sniffing of crown cork closures. What is the ritual of wine pouring: a charming, time-honored component of the overall wine drinking experience, or living testimony to the inability of a winemaker to achieve quality control? Or both?

Another e-mail I received recently was from a man who decried my insistence on preaching the gospel of consistency for beer. Why, he wanted to know, was it so important for every batch of beer to be so consistent. He liked surprises. My retort was that he would not be happy if a batch of anesthetic was out of specification and didn't quell the pain if he was having a surgical procedure. He would equally think it distinctly unacceptable if on opening the can he found his tomato soup to be blue. He would also be a tad miffed to find that gallons of gasoline varied in their ability to support the smooth running of his automobile. To me, the same principle should apply to wine as it most certainly does to beer: have your expectations fulfilled.

A fundamental difference between the production of beer and wine concerns the juxtaposition of raw materials and production process. It is customary for a winery to be built adjacent to the vineyard. Thus, there is a far closer connection between ground and glass for wine, the wine company being at once viticulturalist and winemaker. By contrast, it is very unusual for the brewery to be located adjacent to the barley crop. Rather, it is the malt house that tends to be located close to agriculture, remembering that barley needs sufficient pre-processing before it is in fit condition to be made into beer. While some brewing companies own their own malt houses and produce some or all of their malt (Anheuser-Busch and Coors are notable examples), by far, the majority of brewers buy their malted grain (and adjuncts) from suppliers.

These differences do not mean that the winemaker somehow has a greater control over the character of their products. The brewer is probably more fastidious, demanding rigorous adherence to quality specifications for barley and the malt derived from it, as we shall see in Chapter 5. What it does mean, though, is that the winemaker tends to have a more substantial investment in land and the cultivation thereof: The establishment of a vineyard represents a considerable outlay. A study from my university in 2004 indicated that the cost of establishing a vineyard in Sonoma County, California, exceeded $16,000 per acre. Bear in mind, too, that it takes some four to five years from first planting for there to be a satisfactory harvest of grapes. The yield per acre is likely to be of the order of five tons. The current price (February 2007) for grapes in Napa County is $3,051 per ton, with a statewide average of $581 per ton.

A ton of grapes will yield approximately 150 gallons (around 750 bottles) of wine of average strength. On this basis, then, we can conclude that the cost of grapes for each 750-mL bottle of California wine would be something between 78 cents and a little over $4, depending on where they are grown. Let's say we are buying a bottle of wine from the Central Valley at a cost of $5 in the supermarket or a bottle from a Napa winery at $20: The grapes amount to a fifth of the purchase cost.

Compare the situation with beer: Assuming that a brewer is making an all-malt product (though the cost of adjuncts is a similar order of magnitude), then they may be purchasing that malt at a cost of some $500 per ton. If they are producing bottles containing twelve fluid ounces of a beer that is 5 percent alcohol by volume, then the cost of malt per bottle is approximately 2.5 cents. Let us say we sell beer at a dollar a bottle, then the order of magnitude of malt cost will be about 2.5 percent. Bear in mind, of course, that a bottle of beer with an alcohol content the equivalent of wine (and they do exist – see Chapter 9) would demand pro rata more malt equivalents per bottle, but the cost of this raw material will never reach that of grapes in the wine. However, we should not forget that the brewer has also to pay for hops and water.

Does this cost differential speak to a reality founded on some supposed superiority of the grape as opposed to the grain, akin to comparing pearls with colored stones? Is it a reflection of bulk agricultural capability, with economies of scale in terms of the growing of vast acreages of barley compared to the "cottage industry" craft of the vineyard? Or is it simply a case that the maltster is obliged through laws of competition and user power to suffer deflated price opportunity while, conversely, grapes are retailed at unrealistically inflated price tags, especially from some regions that don't genuinely have that degree of superior quality?

Let us not forget that a major brewing company will be producing beer 24 hours a day for 365 days of the year. There is no concept of a crush for them: all hell let loose for a few weeks after the grapes enter the winery. It is always a mystery to me what those wine guys do for the rest of the year.

But here is the difference: the brewer genuinely sweats his or her assets, whereas the winemaker basically uses the vessels once per annum and will tend to have tanks filled with maturing product for far longer than will the brewer, even those advocating prolonged maturation periods. It truly does not make sense to have a product sitting in storage unless there are genuine changes taking place in the liquid that render the product, be it wine or beer, of superior quality as perceived by the consumer. I would contend that in many instances the storage is a marketing issue, allowing a tale to be told.

2. A Brief History of Wine

*Noah was the first tiller of the soil. He planted a vineyard; and he
drank of the wine, and became drunk . . .*
GENESIS 9:20

*To quench Noah's thirst God created the vine and revealed to him
the means of converting its fruit into wine.*
BENJAMIN FRANKLIN

I T IS PROBABLE THAT WINE WAS NOT THE FIRST ALCOHOLIC BEVE-
rage enjoyed on this planet. Grain was a cultivated crop before
grapes, and the work of the bee led to honey in the ancient forests
at a very early stage. And so, the earliest beers and meads almost
certainly pre-date wine.

Others suggest, however, that wine must have preceded beer
because it is rather easier to make than is beer – a case of just
crushing the grapes and allowing adventitious organisms on the
surface of the berries to do their own thing. As a colleague of mine
says, if you tread grapes you get wine; if you tread grain you get sore
feet.

I facetiously tell my students that Jesus performed the miracle of converting water into wine because doing the trick of water to beer was far too technically demanding. The winemakers remind me that it was his first miracle.

Grain needs some degree of processing before it is ready for brewing, and yeast is not a native inhabitant of cereals and so would either need to be added to make beer or be present as a contaminant in a vessel. The nature of yeast was of course wholly unknown in those far-off days and was still argued over more than 7,500 years later – but the ancients learned empirically that the addition of fruit to the brew would trigger the fermentation process (or whatever they called it in those days), as would the transferring of the brew to containers that had previously been used to store fruit. More importantly, they realized that mixing a little of the old brew with a new one would kick the process into life. Some people refer to this by the somewhat unpleasant term "back slopping." More about this in Chapter 3.

The early grapes grew wild in Eurasia, Asia, and North America and were likely *V. vinifera sylvestris*. They grew either on the ground or by twisting around the trunks and branches of trees. For climatic reasons, the most suitable grapes for converting into wine come from the latitudes between 30 and 50 degrees North and South, and it was in these regions that the first wines were produced.

It is claimed by many that grapes and cereals triggered the lifestyle transition from a nomadic existence to one based on static communities. The ancients must have realized the merits and necessity of staying put while the best grapes and cereals were allowed to grow, prior to processing into wine and beer in increasingly controlled processes.

These civilizations were located in the Fertile Crescent, the region in the Middle East that originally incorporated Ancient Egypt, the Levant, and Mesopotamia. This vast area sweeps from the eastern Mediterranean, embracing the Tigris and Euphrates rivers, and nowadays comprises Syria, Iraq, and Lebanon. Vines were certainly cultivated by man in the Middle East before 4000 BCE, but there is now evidence that controlled growth of vines dates as far back as 6000 BCE in Georgia. With a good degree of certainty, we can date vine cultivation to between 5400 and 5000 BCE in the Northern Zagros Mountains.

Technology would have developed substantially by that time. There was likely a selection of vines with the biggest grapes to afford the highest yield, together with a realization that some berries afforded superior wine to others.

The first vessels were based on animal skins, but the inability to reuse these led to the advent of pottery containers, which were fired to restrict gas permeability and possessed narrow necks that could easily be sealed with pitch and clay. Through trial and error, the early vintners learned that contact of the wine with air under warm conditions rapidly leads to spoilage and that storing containers on their sides further reduced deterioration by lessening the rate of air ingress. The jars were sunk into the ground to keep them cool.

Much of our current understanding of the ancient history of wine stems from the study of residues in earthenware jars, notably the presence of tartaric acid, a sure marker for grape-based products.

The culture and techniques of wine passed from culture to culture as a partner to colonization. Thus, the technology was passed on to the Egyptians, and then to the Cretans, Greeks, Romans, and, in turn, the peoples who inhabited what are now Italy, Spain, France, and Germany. It was ever a marketable commodity. In the older civilizations, wine tended to be the drink for the upper classes, with beer for the proletariat. Alcohol was very much a daily dietary staple and a key component of the nutritional intake. Wine also became associated with religious and secular events, in both pre- and post-Christian eras. The majority of wines were red, and some would have it that there was perceived symbolism with blood.

Southern Mesopotamia, embracing Sumeria and Babylon, had a climate less than ideal for the growing of grapes suitable for converting into wine; hence, the region was in truth beer country. The wine was imported (at cost) for religious purposes and to satisfy the upper classes.

At first Egypt, too, was very much a nation where beer was prized, but grape fermentation developed around 3000–2500 BCE, albeit with the wine being four times more expensive than the beer.

Wine has always tended to be much more alcoholic than beer, making for a longer shelf life. It was long ago realized that smaller volumes of wine have a bigger impact on the senses and this, together with its sweeter and more luscious flavor, led to greater favor, taking the drinker away from the challenges of the day. The wine of old would have been quite nutritious of itself (no clarification in those days so lots of vitamin-bearing yeast), and thus aided the digestion of other foodstuffs, leading to a healthier lifestyle.

Returning to the theme of redness, the Egyptians also compared the hue of the wine with that of the Nile arising each year from one of its tributaries. The "rebirth" of this water and annual reemergence of the vines symbolized fertility.

Wine also featured prominently in the ancient pharmacopoeias, as a sedative, a diuretic, and a treatment for anemia as well as an anesthetic, appetite stimulant, antiseptic, and anti-diarrheal. It was also used in poultices. In Egypt, other medicines were mixed into wine and beer, which formed the base for medications. One cure for epilepsy called for the testicles of an ass to be finely ground in wine. I, for one, would delight in hearing a latter-day wine writer attempt to articulate the aromatic nuances of that particular beverage.

Successively, as the wine world expanded, so did the advent of technology. Nearly 2,000 years ago, Pliny was classifying grapes according to soil type, ability to withstand disease, time to ripen, color, and the wines that could be produced from them. The Greeks grew vines on trellises as opposed to on trees, as was practiced previously. The Romans were familiar with pruning and, in the first century CE, replaced the Greek pottery amphorae with barrels. The treading of grapes was supplemented by pressing – a Cretan innovation employing boards weighed down by rocks. Later, the Greeks and Romans employed heavy beams and screw mechanisms for that purpose, although it was still recognized that "traditional" treading yielded wines with higher quality and longer shelf life

It seems that most of the early wines were red, with the grapes left in situ through fermentation. White wines were in limited availability, albeit prized in Greece and Rome. They were almost certainly amber or brown in color as a consequence of warm storage and air ingress.

The ancient wines were likely flavored by herbs and spices, either directly or by aroma transfer from plants such as lavender and thyme grown alongside the vines. Furthermore, the resin and pitch from the fermentation and storage vessels afforded flavors of the type to be found nowadays in retsina. The Greeks were hardly reticent in their use of flavorings: Another favored addition was grated goat's milk cheese.

The Greeks relished their drinking sessions, which were known as symposia. Essentially they were orgies, making a modern-day Friday night in a London pub a church picnic by comparison. The Roman equivalent of the symposium was the convivium. While the upper echelons were partaking of their wine, the poorer people drank posca, a soured wine diluted with water and lora, which was made by soaking the grape press residues in water.

Greek and Roman wines were sweet. The sugar content was maximized by the drying of grapes before fermentation and the addition of boiled must and honey post-fermentation. In turn, the sugariness and off-flavors were tempered by diluting the wine with seawater.

No matter how appealing (or otherwise) this sounds to us nowadays, the merits of wine were far from denied in those days. Thus, Plato opined that folk older than 40 could drink freely to "relieve the desiccation of old age."

Indeed, a word search of the Bible leads to more "hits" for the grapevine than any other plant. There is a mixed press, however, for the drink. In his epistle to Timothy, Paul urges "have a little wine for the sake of your digestion." Hosea, meanwhile, recommends maturity in the product, saying "new wine addles the wits." And, as we have seen, Christ's first miracle was turning water into wine at Cana. It was the best wine, too – not the rubbish held back for when folks lose their discriminatory skills. For as it says in John 2:10: "Every man at the beginning doth set forth good wine; and when men have well drunk, then that which is worse."

And so by the earliest years of the spread of Christianity worldwide, wine was in production in such unlikely locales as England. The Romans retained a superiority complex, feeling that their wine-drinking tendencies set them on a higher pedestal than the Germanic tribes and the other cultures encountered, who drank beer. And thus we have Edward Gibbons in his description of the decline of the Roman Empire stating that the Germans were "immoderately addicted to strong beer...extracted with very little art from wheat or barley and corrupted...into a certain semblance of wine." For their part, macho Germans felt that men drinking wine would adopt feminine characteristics.

It was the Church that played an instrumental role in spreading viticulture and enology practices throughout Europe. In England, then, the Domesday Book of 1086 recorded 42 vineyards, but as the thirteenth century drew to an end, there were 1,300 vineyards. Gradually, these were replaced by other crops: Fundamentally, the English climate is not suited to growing grapes for wine and the imported wine was clearly superior to anything that could be fermented locally.

Through the eleventh to fifteenth centuries, the demand for wine in England grew hugely. Wines came into England from Bordeaux and Anjou. This later emerged into a demand for sweeter, more alcoholic products, with a longer shelf life, from the Mediterranean. The loss of Gascony to France in 1453 led to a decline in wine from the victorious country and opportunities for Spain. Thus emerged Spanish "sack" – sweet wine – a word derived from *sacar*, meaning "to draw out" – that is, a wine suitable for exporting. The Spanish wine industry grew apace in the sixteenth century. The trade of port to England grew in late seventeenth century when political pressure led to the banning of French wine imports.

England was emerging as the hub of the world wine industry despite being a location not suited of itself to wines of excellence. Here, too, was the original seat of wine literature.

Yet, in England, as in many countries, beer remained the chief alcoholic beverage, and would continue as such throughout the sixteenth to eighteenth centuries. Beer was far cheaper than imported wines. And so, in ecclesiastic climes, the Abbot sipped his wine while the brothers quaffed ale, taking their wine only on festival days. Beer was less dependent on geography for its main raw material than was wine.

The year 1662 saw one of the earliest attempts by winemakers seeking to gain an advantage over beer with the authorities in Bordeaux banning the brewing of beer as it constituted a threat to the local wine business.

Wine was facing threats from more than beer alone. The Dutch emerged as a major importer and reexporter of brandy (Brandewijn = burnt wine). Vodka, whiskey, rum, and gin all emerged to compete for what the twentieth-century alcoholic drinks analyst company Canadean calls a "Share of Throat." Quite simply, these beverages, to which we might add schnapps, are somewhat less dependent than is wine on the vagaries of agriculture and climate for their successful production. Distillation can be a great leveler.

The march of innovation continued apace. Thus, the best wines were being filled into glass bottles by the sixteenth century, with the current familiar bottle style emerging in England in the 1630s following improvements in furnace technology. Wine glasses replaced the hitherto customary pottery, leather, wood, and silver. Likewise, cork replaced leather, wood, and textiles for sealing bottles in the seventeenth century, but of necessity had to follow the invention of the corkscrew. Now we could have champagne.

In the eighteenth century, the burning of sulfur in casks to fumigate them was a direct forerunner of the longstanding presence and importance of sulfur dioxide in wine. The clearer wines were considered to be more appealing, especially now that the wine was being sold and drunk in glass, and so emerged the practice of fining, by agents such as egg white and isinglass. With the arrival in Europe of sugar shipped back from empires, it was possible to enrich musts, leading to more alcohol and more consistency. The addition of sugar is referred to as Chaptalization, after Jean-Antoine-Claude Chaptal, though actually he probably didn't invent the process.

Around 1000 CE, Leif Ericsson sailed west from Greenland and landed at a place ostensibly abundant with grapes, which he named Vinland. We now believe it to have been what is now Newfoundland. Some doubt that he encountered grapes, suggesting that the fruit was far more likely to have been cranberries.

With more certainty can we state that Spanish colonizers took wine to South America and to Central America. The first winemakers in North America were probably Spanish settlers in what is now South Carolina. There were attempts to grow grapes and make wine in the early English colonies in Virginia, but the settlers quickly realized that tobacco was a more successful crop.

It was from California, of course, that the North American wine industry really established itself as a world player, as will be discussed in Chapter 8. The irony is that some of the first winemakers pitched up on the Pacific Coast as a consequence of America exporting a dreaded grape disease to Europe. In 1870, certain American varieties were shipped to botanical gardens in France and England. Unfortunately, the grapes harbored the root louse Phylloxera. The American vines were resistant to it, whereas the European vines (*Vitis vinifera*) were not. The host vineyards were devastated and remained so for thirty years. The response was to bring into France the vigorously growing, phylloxera-resistant American rootstock and graft the favored European scion to it. Unfortunately, American varieties also delivered an undesirable foxy flavor.

The first vines in South Africa were planted in the Cape by Jan van Riebeeck in 1655, with the first wine emerging four years later. Vines were carried on board the first ships into New South Wales in 1788 and the first wine was enjoyed there in 1795.

3. A Brief History of Beer

A high and mighty liquor, made of barley and water.
JULIUS CAESAR

THE HISTORY OF BEER IS CERTAINLY NO SHORTER OR LESS ILLUS-
trious than that of wine or, indeed, that of other beverages that
evolved at a similar time by the "spontaneous" fermentation of a
diversity of materials: from moistened grain (beer), through grapes
(wine), to honey (mead), and milk (kvass).

In all instances, these products, with their hedonic attributes and
the entry they allowed to an altered state of consciousness, had
a profound social and economic impact. Furthermore, as we saw
in Chapter 2, for wine and as is almost certainly more likely the
case for beer, they featured as prime causative factors in the advent
of static societies. They were a fundamental part of the diet and
were undoubtedly recognized as being altogether healthier to drink
than water alone. Alcohol is a great killer of pathogenic microbes
but more than that, a professor of anthropology at Emory University, George Armelagos, found evidence for the bacterium Strepto-
mycedes in relics from ancient Sudan and has suggested that beer at

the times was likely a significant source of the "natural" antibiotic tetracycline. Beer is almost a wonder food: nutritionally enriched, hedonistically satisfying, and medically protective.

Climate, of course, played a substantial role in dictating the beverage of choice. In northern and western Europe, then, it was probably honey that constituted the first alcoholic beverage base, before the days of cultivation of the cereal grasses. In more southerly Europe and the Eastern Mediterranean, the grape was clearly the more important source, though we suppose that the date palm predated the grape. It is likely, too, that Kvass was the forerunner of beer.

Grain has formed a staple part of the human diet for thousands of years. Spontaneous germination of grain by accidental wetting during rainfall led the early consumers to realize that it was softer, somewhat more tasty (less astringent, more sweet), and easier to digest after this sprouting. People fared better on it because the nutrients were more readily digestible. They would also have ascertained that a subsequent drying of the grain further enhances its flavor.

At some point, a porridge of the sprouted grain was likely spontaneously seeded with yeast – perhaps it was in grain stored in a container that had previously held fruit. The consumption of the gruel, that now would contain alcohol, no doubt had a pleasing impact on the consumer – the forerunner of the modern beer had been discovered.

The earliest human civilizations were in locales where the chief grain crops were grown: West and Central Eurasia, East Asia, and Central America. It may indeed be the case that beer evolved independently in different cultures around the world. Yet it was certainly in the Near East of some 8,000 years ago that the barley and wheat growing wild were domesticated and the processes of bread making and brewing developed. Most likely it was baking that emerged first for, as we shall see, it was grain that was converted into loaves that constituted the raw material for some of the earliest brewing processes.

The first acclaimed seat of brewing was Mesopotamia (the latter-day Iraq), with a plethora of archaeological and art evidence informing us of the important role that beer had to play in society. It is now believed that brewing within Mesopotamia can be traced to Sumeria and later to Babylonia (capital Babylon).

The first evidence of people enjoying beer is to be found at Tepe Gawra, in modern-day Northern Iraq, with seals from 4000 BCE depicting the drinking of beer through straws from a huge jar.

Kiln-dried malt in Sumeria was pounded and sieved, and the crushed malts stored or made immediately into doughs for oven baking. The small loaves were known as bappir. Various aromatics may have been incorporated into the loaves. Emmer may have been used: This was a primitive type of wheat that differed from modern cultivars by retaining a hull.

The loaves were crumbled, mixed with water, and heated, that is, they were mashed. After cooling, date juice or maybe honey may have been added. Some say that the mix was boiled, but there is little evidence for this. It is likely that fermentation was triggered by yeast naturally seeded from additions of fruit, but it was likely soon realized that a more reliable trigger to fermentation was the addition of a portion of the previous brew.

After fermentation, the mixture was transferred to another vessel to clarify and then the liquid was drawn off and poured into jars for storage or transportation. Despite the clarification stage, the finished product was far from "bright," hence, the preference for drinking through straws, which were made literally of straw or, for the successively higher walks of life, copper, silver, or gold.

These ancient practices were captured in the Hymn to Ninkasi (1800 BCE):

Borne of the flowing water,
Tenderly cared for by the Ninhursag,
Borne of the flowing water,
Tenderly cared for by the Ninhursag,

Having founded your town by the sacred lake,
She finished its great walls for you,
Ninkasi, having founded your town by the sacred lake,
She finished its walls for you,

Your father is Enki, Lord Nidimmud,
Your mother is Ninti, the queen of the sacred lake.
Ninkasi, your father is Enki, Lord Nidimmud,
Your mother is Ninti, the queen of the sacred lake.

You are the one who handles the dough [and] with a big shovel,
Mixing in a pit, the bappir with sweet aromatics,
Ninkasi, you are the one who handles the dough [and] with a big shovel,
Mixing in a pit, the bappir with [date] – honey,

You are the one who bakes the bappir in the big oven,
Puts in order the piles of hulled grains,
Ninkasi, you are the one who bakes the bappir in the big oven,
Puts in order the piles of hulled grains,

You are the one who waters the malt set on the ground,
The noble dogs keep away even the potentates,
Ninkasi, you are the one who waters the malt set on the ground,
The noble dogs keep away even the potentates,

You are the one who soaks the malt in a jar,
The waves rise, the waves fall.
Ninkasi, you are the one who soaks the malt in a jar,
The waves rise, the waves fall.

You are the one who spreads the cooked mash on large reed mats,
Coolness overcomes,
Ninkasi, you are the one who spreads the cooked mash on large reed mats,
Coolness overcomes,

You are the one who holds with both hands the great sweet wort,
Brewing [it] with honey [and] wine
(You the sweet wort to the vessel)
Ninkasi, (…)
(You the sweet wort to the vessel)

The filtering vat, which makes a pleasant sound,
You place appropriately on a large collector vat.
Ninkasi, the filtering vat, which makes a pleasant sound,
You place appropriately on a large collector vat.

When you pour out the filtered beer of the collector vat,
It is [like] the onrush of Tigris and Euphrates.
Ninkasi, you are the one who pours out the filtered beer of the collector
    vat,
It is [like] the onrush of Tigris and Euphrates.

The legendary brewer Fritz Maytag, president of the Anchor
Brewing Company in San Francisco, worked with the historian
Solomon Katz to recreate a brew according to the precepts of
Ninkasi and available archaeological evidence, right down to the
sipping of beer through straws at a banquet in the city by the bay.
There is actually no reference to malting in the Hymn, which leads
some to suppose that the grain was not germinated and the loaves
baked from a meal of raw barley and emmer. A range of flavorings
is believed to have been used for the beer in Sumeria, *inter alia*
lupin, safflower, mandrake, grape pips, dates, coriander, fenugreek,
and bitter orange peel.

From Mesopotamia, the techniques of brewing passed to the
Egyptians on both banks of the Nile. Beer sediments have been
found in jars dating from the pre-Dynastic era of 5500–3100 BCE in

Egypt. There are written records from 3100–2686 BCE in the early Dynastic period that prove beer to be a fundamental part of their culture; indeed, some claim that bread, beer, and onions were staples of the diet by 6000 BCE.

The beer would not have been unduly strong and, in view of its short shelf-life, would have been brewed, as well as drunk, daily by all echelons of society. It was a fundamental part of religious festivals – an offering to the Gods – and integral to the national economy. In the pre-Dynastic era, grain in the form of beer or bread was how wages were paid. In later Ptolemaic times, taxes were levied on beer, and governments have realized ever since how alcoholic beverages can constitute a powerful source of income to the exchequer.

Ancient Egyptian brewing techniques were similar to those developed in Sumeria and the drink Bouza, which to this day is produced in Egypt, was made in such a way. Bread made from germinated grain with light baking was crumbled and mixed with water, strained, and left to ferment. The liquid was enriched in sugars and further flavored by the addition of dates. Some contend that lupin and mandrake were included, and it is likely that "back slopping" was employed, whereby a small portion of the previous brew was added to seed and triggered the new fermentation. Brewing was fundamentally women's work, whereas the later wine making was the preserve of men.

The Egyptians exported their brewing know-how to the Greeks, who in turn transmitted the skills to Gaul, Spain, and the eastern Adriatic. However, the Greeks and Romans were wine lovers. Wine there, later in Egypt, and nowadays perhaps globally, had snob value. Grapes were probably up to ten times more expensive than barley, leading to the inherent (yet inevitably inaccurate) supposition that a product derived from a more valuable commodity must of itself be of higher quality. Various people extolled the virtues of wine and decried beer. There were even those prepared to attest that wine and beer had different impacts on the body. For example, Aristotle proclaimed: "Men who have been intoxicated with wine fall down face foremost, whereas they who have drunk barley beer lie out-stretched on their backs; for wine makes one top-heavy, but beer stupefies."

Wine was substantially more alcoholic than beer and so rather more intoxicating – however, it also enjoyed a longer shelf life and did not turn to sour vinegar as quickly as beer.

As I have alluded to previously, alcoholic beverages may have evolved quite independently in different parts of the world. Thus, in Central and South America, forerunners of drinks such as "chichi"(based on corn) emerged, with the initial stage of prepara-tion being the chewing of the grain in which salivary enzymes were able to commence the production of fermentable sugars from the starch, and the next stage being the heaving by all the chewers of the contents of their mouths into a communal bucket for fermentation.

To return east, we actually find no word for beer in the Bible, the Hebrew *birah* coming into play later than the Biblical texts were laid down. Wine (*wayin*), of course, gets a mixed press in both testa-ments. There are fervent neo-Prohibitionist zealots who insist that reference to wine in the Bible is to unfermented grape juice but I for one find it hard to contemplate the ancients having such con-trol over the natural fermentation caused by adventitious yeasts that the grape juice (and pretty much any other liquid sugary source) would not quickly turn to alcohol and be preferred for that very reason.

Indeed, there are those who are quite prepared to argue that beer really is in the Bible. For example, James Death proposed in

1887 that the passage from Exodus, Chapter 12, verse 19, which reads: "Seven days shall there be no leaven found in your houses: for whosoever eat that which is leavened, even that soul shall be cut off from the congregation of Israel, whether he be a stranger, or born in the land" actually referred to the consumption of a bread (unleavened – compare the bappir of ancient Sumeria) as opposed to beer (leavened – compare bouza).

Indeed, Death was moved to write (in the face of what we encountered earlier): "I adduce reasons to show that the manufacture of beer was the earliest art of primitive man; an art exceeding in antiquity that of the potter or of the wine-maker, and certainly that of the baker."

The Phoenicians probably played a significant part in spreading beer and brewing across the world, even as far as Britain, and may even have brewed beer on board their ships.

However, the credit for introducing brewing to the British Isles is usually given to the Celts. There are those who doubt that the Celts could have gained their mastery over brewing from Egypt, but certainly the northern Celtic tribes (Gauls and Belgiae) recognized beer as being superior to wine, the latitudes being too northerly for the growth of suitable grapes. They enjoyed mead, too. The early cereal used to make beer in Northern Europe was likely emmer, there being no hulled barleys bred at the time.

A further candidate for introducing beer to the British Isles is the Beaker people. Some of the earliest evidence is from pot shards located near Glenrothes in Scotland, which is evidence for human activity in the fourth to second millennia BCE, that is, from the Neolithic to the Bronze Ages. Herein lies evidence for brewing, with remnants among a range of plant tissues of barley, oats, and burned cereal mash. There was also meadowsweet, flax, cow parsley, and henbane. And thus it has been claimed that it was the Beaker people who established barley as a crop in the British Isles. The beverage produced therefrom would naturally have been consumed ritualistically from a beaker, and thus the name "Beaker people." Grapes did grow in the region, but had low sugar content. And as the grain was more abundant than the honey, beer was probably more popular than mead, which was likely reserved for special occasions.

The Celts came to Britain 1,500 years after the Beaker people, circa 500 BCE. They had the technology to work wood and metal (Iron Age) and thus the barrel evolved, beer's equivalent of the amphora.

Actually, some of those ancient beverages may not have been based on a single raw material. For instance, evidence from what is now Denmark suggests that honey, cranberries, and cereal grains were combined for the production of certain alcoholic drinks.

The Romans probably grew barley throughout their empire for the production of bread and beer. Indeed, unearthed Roman coins depict barley. Of course, wine was the beverage of choice for the Romans. It was that nation who introduced the first taverns (*tabernae*), located strategically alongside their newly constructed roads.

As we move to the Anglo-Saxon era we discover that beer was consumed copiously. We find, too, a key role for the monasteries in developing the controlled art of brewing, and ale was a major component of the diet in both secular and non-secular climes. It was quickly realized that ale was substantially safer to drink than the water.

Various words have been handed down from the era, the meaning of which have only latterly been unraveled. Although *bior* sounds logically to be the forerunner of the latter day product, it may actu-ally have been used to describe either a honey-based beverage or a cider. *Ealu* would have been less alcoholic and was likely the grain-based product (leading in due course to the word *ale*). *Win* and *meordu* were the early wines and mead. Scholars agree that wine was at the top of the prestige pecking order, followed by bior, meordu, and lastly ealu.

One of the earliest champions of sound brewing practice was Charlemagne and wherever his empire spread there sprang up monasteries as the seats of brewing excellence. It was in these monasteries that experimentation began on the use in beer of pro-prietary blends of herbs and spices known as *gruit*. The use of gruit, which included components such as mugwort, sweet gale, heather, yarrow, ginger, caraway, juniper, nutmeg, cinnamon, and juniper, perhaps triggered the boiling of the extracts, so as to enhance the extraction of flavors. Prior to this, it may well have been that all materials were added to a single vessel before mixing with warm water, prior to the ensuing fermentation.

It was probably with the later introduction of hops in place of gruit that the mashing and boiling stages were separated. It was also likely realized that you could collect liquids of different strengths, which could be boiled and fermented separately to give products of different potencies.

Royalty have long had an interest in beer. In the realm of King Edgar, Archbishop Dunstan decreed that pins or nails should be put into drinking cups to regulate how deep a draught the drinker could enjoy with each sip. This inevitably led to drinking challenges, in which consumers could see how quickly they reached the next marker, in other words "taking down a peg."

Throughout the Norman Conquest, the English remained faith-ful to their ale, which by that time was likely a barley malt- and oat-based beverage, albeit one as yet lacking in hops. Spices were used to add flavor.

One of the quality control innovations sponsored by William the Conqueror was the employment of ale conners, men in leather breeches who would visit breweries and pour the ale onto wooden seats before sitting in it. If they stuck to the seat it meant that the sugar had not been efficiently converted into alcohol and woe betide the brewer.

Through the Assize of Ale and Bread in 1267, Henry III extended the powers of the ale conners such that they were empowered to alter the price of the ale depending on their judgment of its quality. The ale conners knew when there was ale to test because the brewer placed an "ale-stake" (a branch or bush) outside the premises when a new batch had been brewed. The concept extended to advertising when ale was available to drink and evolved to the hanging of a metal hoop, later by objects displayed in the hoop, and finally to a picture of the object, just as today we have pub signs. The use of pictures allowed the same name for far more premises (there could only have been one King's Head before the use of artwork) and allowed a greater diversity of names, for example, it would not hitherto have been feasible to have the Coach and Horses.

One notable example of the pride that the English felt over their beer and its assumed superiority over wine came with Thomas Becket's journey to France in 1158. Henry II was in search of an alliance with King Louis, which would be served by the betrothal of his son to the French king's daughter. Becket was dispatched to Paris, taking with him two horse-drawn wagons loaded with ale. One observer relished the sight of "ale, decocted from choice fat grain, as a gift for the French who wondered at such an invention – a drink most wholesome, clear of all dregs, rivaling wine in color, and surpassing it in savor."

This was at a time when monks were consuming a gallon per day of full strength ale and perhaps a further gallon of weaker ale. Within the monasteries, the controlled and revered brewing practices were being maintained. Outside the monasteries, the tradition of women *brewsters* producing beer in the home continued. In some parts of England, as many as 75 percent of households were brewing their own ale.

The Magna Carta of 1215 did not ignore alcohol. Rather, it insisted upon standardized quantification, with clause 35 stipulating, "Let there be one measure of wine throughout our whole realm; and one measure of ale."

The first body addressing areas of shared interest in brewing was the Brewers Company in London of 1312. Similar groups grew up all over the country, and in some locales, they regulated how much could be brewed and set the price of ale.

The English were certain that their superiority over the French in the Hundred Years War was on account of the fact that they drank ale and not the "small sour swish-swash of the poorer vintages of France." It has long been held that the first hopped beer was landed in Winchelsea in 1400 from Holland. It was intended for the Dutch merchants working in England who didn't like sweet English ale. However, in 1971, a boat was discovered in Graveney marshes near Whitstable, a vessel that may have dated to the year 949, and among the plant remains on board were hops.

Hops had been used on the continent of Europe for many years, with records showing that they were grown in 736 in the Hallertau region of Bavaria. The first reference to beer brewed with hops is from 822, and a monastery on the River Weser in Germany. Hops grew in abundance on the flood plains of rivers and on the edge of forests. They were cultivated from mid-ninth century onwards, notably in Bohemia, Slovenia, and France, as well as Bavaria. The Benedictines were known for their growing of hops.

And yet some historians believe the word hop comes from the Slav *hmelj*, which in turn originated in Finland. The claim is that there is a mention of hops in the Finnish saga, *The Kalevala*, which may be 3,000 years old.

Rune XX includes the following:

> Louhi, hostess of Pohyola,
> Hastens to the hall and court-room,
> In the centre speaks as follows:
> "Whence indeed will come the liquor,
> Who will brew me beer from barley,
> Who will make the mead abundant,
> For the people of the Northland,
> Coming to my daughter's marriage,
> To her drinking-feast and nuptials?
> Cannot comprehend the malting,
> Never have I learned the secret,
> Nor the origin of brewing."

> Spake an old man from his corner:
> "Beer arises from the barley,
> Comes from barley, hops, and water,
> And the fire gives no assistance.
> Hop-vine was the son of Remu,
> Small the seed in earth was planted,
> Cultivated in the loose soil,
> Scattered like the evil serpents
> On the brink of Kalew-waters,
> On the Osmo-fields and borders.
> There the young plant grew and flourished,
> There arose the climbing hop-vine,
> Clinging to the rocks and alders.

> "Man of good-luck sowed the barley
> On the Osmo hills and lowlands,

And the barley grew and flourished,
Grew and spread in rich abundance,
Fed upon the air and water,
On the Osmo plains and highlands,
On the fields of Kalew-heroes.

"Time had travelled little distance,
Ere the hops in trees were humming,
Barley in the fields was singing,
And from Kalew's well the water,
This the language of the trio:
'Let us join our triple forces,
Join to each the other's powers;
Sad alone to live and struggle,
Little use in working singly,
Better we should toil together.'

"Osmotar, the beer-preparer,
Brewer of the drink refreshing,
Takes the golden grains of barley,
Taking six of barley-kernels,
Taking seven tips of hop-fruit,
Filling seven cups with water,
On the fire she sets the caldron,
Boils the barley, hops, and water,
Lets them steep, and seethe, and bubble
Brewing thus the beer delicious,
In the hottest days of summer,
On the foggy promontory,
On the island forest-covered;
Poured it into birch-wood barrels,
Into hogsheads made of oak-wood.

"Thus did Osmotar of Kalew
Brew together hops and barley,
Could not generate the ferment.
Thinking long and long debating,
Thus she spake in troubled accents:
'What will bring the effervescence,

Who will add the needed factor,
That the beer may foam and sparkle,
May ferment and be delightful?'

"Osmotar, the beer-preparer,
Kapo, brewer of the liquor,
Deeply thought and long considered:
'What will bring the effervescence,
Who will lend me aid efficient,
That the beer may foam and sparkle,
May ferment and be refreshing?'

"Osmotar, the beer-preparer,
Brewer of the beer of barley,
Used the beer-foam as a ferment;
But it brought no effervescence,
Did not make the liquor sparkle.

"Osmotar, the beer-preparer,
Thought again, and long debated:
'Who or what will bring the ferment,
That my beer may not be lifeless?'

"Osmotar, the beer-preparer,
Placed the honey in the liquor;
Kapo mixed the beer and honey,
And the wedding-beer fermented;
Rose the live beer upward, upward,
From the bottom of the vessels,
Upward in the tubs of birch-wood,
Foaming higher, higher, higher,
Till it touched the oaken handles,
Overflowing all the caldrons;
To the ground it foamed and sparkled,
Sank away in sand and gravel.

"Time had gone but little distance,
Scarce a moment had passed over,
Ere the heroes came in numbers
To the foaming beer of Northland,
Rushed to drink the sparkling liquor.

Ere all others Lemminkainen
Drank, and grew intoxicated
On the beer of Osmo's daughter,
On the honey-drink of Kalew.

"Osmotar, the beer-preparer,
Kapo, brewer of the barley,
Spake these words in saddened accents:
'Woe is me, my life hard-fated,
Badly have I brewed the liquor,
Have not brewed the beer in wisdom,
Will not live within its vessels,
Overflows and fills Pohyola!'

"From a tree-top sings the redbreast,
From the aspen calls the robin:
'Do not grieve, thy beer is worthy,
Put it into oaken vessels,
Into strong and willing barrels
Firmly bound with hoops of copper.'

"Thus was brewed the beer or Northland,
At the hands of Osmo's daughter;
This the origin of brewing
Beer from Kalew-hops and barley;
Great indeed the reputation
Of the ancient beer of Kalew,
Said to make the feeble hardy,
Famed to dry the tears of women,
Famed to cheer the broken-hearted,
Make the aged young and supple,
Make the timid brave and mighty,
Make the brave men ever braver,
Fill the heart with joy and gladness,
Fill the mind with wisdom-sayings,
Fill the tongue with ancient legends,
Only makes the fool more foolish."

When the hostess of Pohyola
Heard how beer was first fermented,
Heard the origin of brewing,

Straightway did she fill with water
Many oaken tubs and barrels;
Filled but half the largest vessels,
Mixed the barley with the water,
Added also hops abundant;
Well she mixed the triple forces
In her tubs of oak and birch-wood,
Heated stones for months succeeding,
Thus to boil the magic mixture,
Steeped it through the days of summer,
Burned the wood of many forests,
Emptied all the springs of Pohya;
Daily did the forests lesson,
And the wells gave up their waters,
Thus to aid the hostess, Louhi,
In the brewing of the liquors,
From the water, hops, and barley,
And from honey of the islands,
For the wedding-feast of Northland,
For Pohyola's great carousal
And rejoicings at the marriage
Of the Maiden of the Rainbow
To the blacksmith, Ilmarinen,
Metal-worker of Wainola.

Kaukomieli gazed and pondered,
Studied long the rising smoke-clouds;
Came not from the heat of battle,
Came not from the shepherd bonfires;
Heard they were the fires of Louhi
Brewing beer in Sariola,
On Pohyola's promontory;
Long and oft looked Lemminkainen,
Strained in eagerness his vision,
Stared, and peered, and thought, and wondered,
Looked abashed and envy-swollen,
"O beloved, second mother,
Northland's well-intentioned hostess,

Brew thy beer of honey-flavor,
Make thy liquors foam and sparkle,
For thy many friends invited,
Brew it well for Lemminkainen,
For his marriage in Pohyola
With the Maiden of the Rainbow."

Finally the beer was ready,
Beverage of noble heroes,
Stored away in casks and barrels,
There to rest awhile in silence,
In the cellars of the Northland,
In the copper-banded vessels,
In the magic oaken hogsheads,
Plugs and faucets made of copper.
Then the hostess of Pohyola
Skilfully prepared the dishes,
Laid them all with careful fingers
In the boiling-pans and kettles,
Ordered countless loaves of barley,
Ordered many liquid dishes,
All the delicacies of Northland,
For the feasting of her people,
For their richest entertainment,
For the nuptial songs and dances,
At the marriage of her daughter
With the blacksmith, Ilmarinen.

When the loaves were baked and ready.
When the dishes all were seasoned,
Time had gone but little distance,
Scarce a moment had passed over,
Ere the beer, in casks imprisoned,
Loudly rapped, and sang, and murmured:
"Come, ye heroes, come and take me,
Come and let me cheer your spirits,
Make you sing the songs of wisdom,
That with honor ye may praise me,
Sing the songs of beer immortal!"

Straightway Louhi sought a minstrel,
Magic bard and artist-singer,
That the beer might well be lauded,
Might be praised in song and honor.
First as bard they brought a salmon,
Also brought a pike from ocean,
But the salmon had no talent,
And the pike had little wisdom;
Teeth of pike and gills of salmon
Were not made for singing legends.

Then again they sought a singer,
Magic minstrel, beer-enchanter,
Thus to praise the drink of heroes,
Sing the songs of joy and gladness;
And a boy was brought for singing;
But the boy had little knowledge,
Could not praise the beer in honor;
Children's tongues are filled with questions,
Children cannot speak in wisdom,
Cannot sing the ancient legends.

Stronger grew the beer imprisoned
In the copper-banded vessels,
Locked behind the copper faucets,
Boiled, and foamed, and sang, and murmured:
"If ye do not bring a singer,
That will sing my worth immortal,
That will sing my praise deserving,
I will burst these bands of copper,
Burst the heads of all these barrels;
Will not serve the best of heroes
Till he sings my many virtues."

Wine may attract rather more literary attention than beer these days, but in the *Hymn to Ninkasi* and *The Kalevala*, there is a rich and ancient pedigree of prose devoted to beer.

Hops have tremendous preservative value and so hopped beer traveled better and had a longer shelf life. Before the advent of hop usage in the United Kingdom, the ale (a term then restricted to unhopped beer) was perforce strong, such that the high alcohol content afforded protection against spoilage. By using hops, the brewer could retail the beer at lower alcohol content – much to the distaste of English traditionalists, some of whom referred to hops as a "wicked and pernicious weed."

The brewers of beer were natives of Holland and Zeeland and were harassed as "foreigners" by the English ale brewers. Henry VIII went so far as to forbid the use of hops in 1530, as they were an affront to "good ole English ale." Indeed, they were deemed Protestant plants, coming as they did from the Low Countries. Two decades later Edward IV repealed the ban and the terms ale and beer rapidly became synonymous. Henry VIII had other impacts on the brewing industry, with his dissolution of the monasteries. The brewing traditions continued in universities and with the brewsters (the alewives) and victualler-brewers (the equivalent of the modern brew pub). The alehouses sold beer only, the taverns also sold wine, while inns provided a bed, too.

The pride that the English felt in their beer was encapsulated in the words of the herald of England, Sir John Coke, who said in 1549 to his French equivalent: "...we have good ale, bere, metheghelen, sydre, and pirry, being more holsome beverages for us than your wynes, which maketh your people drunken, also prone and apte to all filthy pleasures and lustes." Doyen of royal beer lovers was Elizabeth I, whose entourage ensured she received brews of excellence to delight her, wherever her travels took her in the realm.

Hops were grown all over England and into Wales by end of six-teenth century. Even so, many flavorings were used, perhaps more than 150 plants, including bay leaves, orrice roots, bog myrtle, yarrow, marsh rosemary, St. Johns wort, juniper, caraway, and worm-wood. Then, as now, there was a far greater diversity in beers, their ingredients and their nature, than there is in wine.

Over in Bavaria, the first move towards restricting the use of unwholesome materials was put in place in 1487 by Duke Albrecht IV, who decreed that barley, water, and hops alone could be used for brewing beer (yeast was, of course, unknown at the time). This translated into the famed Reinheitsgebot of 23 April 1516, encrypted by Duke Wilhelm.

As the industrial revolution advanced, the brewing industry came to be dominated by men. So-called "common brewers" became larger, as brewing shifted away from the home. The factories employed the majority of urban people; sizeable brewing companies provided the beer that had hitherto been the brewsters' preserve. Furthermore, there were changes in the character of beer introduced directly as a result of industry. The shortage of wood demanded for building ships led to a replacement of wood by coal for kilning, with a direct impact on the flavors imparted to the malt (see the story of Porter in Chapter 9).

The challenge to brewers from other drinks continued. Dubious taxation laws meant that gin attracted far less duty than did beer, leading directly to a plethora of distilleries producing largely unregu-lated and dangerous gin products and serious problems for the health of Londoners. The first coffee house was opened in Oxford in 1650, adding coffee to chocolate as a popular drink. Gin was taken in these houses. Tea came later.

In 1673, the brewers championed a petition to Parliament that claimed tea, coffee, brandy, and mumm (an unhopped ale from Braunschweig – sweet, sugary, based on wheat malt and flavored with diverse aromatics) were "detrimental to bodily health... instead of our national beverage, sound barley beer."

By the mid- nineteenth century there were three strengths of beer in Britain, based on successive mashing of the same batch of malt. The runnings from the first mash were naturally the strongest, and there followed an addition of more water and the ensuing wort was of intermediate strength. The runnings from the last infusion were known as small beer or, less charitably and owing to their relative inconsistency, as "rot gut."

Scientifically, brewing has always been more innovative than has wine making. And so, the thermometer was first used in brewing by James Baverstock, Jr., in 1760, prior to which brewers had employed the "rule of thumb" – water was ready for use in mashing if one's hand could just stand the heat or if your face was reflected in the water. Baverstock's father had experimented with a saccharometer (hydrometer) in 1768, but the use of that instrument was particularly championed by John Richardson in 1785.

The first brewer to install a steam engine was Cook and Company at Stratford-Le-Bow in 1777. There was now little to arrest the huge growth in beer volumes, except perhaps space in the inner cities, causing breweries to grow upwards in tower formats, which in any event allowed the use of gravity for the natural flow of liquid streams.

The forerunner of the modern Napa wine trail was surely the London brewery scene, and in the nineteenth century their huge vessels constituted a significant tourist attraction. With such large vessels, it was ever a challenge to maintain temperature control.

Brewing at first was restricted to the cooler months of October through April. However, the introduction of the cooling coil by Long in 1790, the forerunner of the heat exchanger of Yandall in 1826, and refrigeration by Linde around 1870, allowed year-round brewing and, therefore, income twelve months in the calendar. In large countries such as the United States, beer could be distributed over vastly greater distances in refrigerated railcars. Breweries could (and did) get bigger and bigger. The nineteenth century also saw the development of sparging processes, while Joseph Bramah developed the beer engine for dispense of cask ale at the end of the eighteenth century.

Taxation of beer was relentless – among the wars being funded on the back of the strength of beer was the American War of Independence. Gradually, a British system based on taxing the raw materials, malt and water, was replaced by one in which duty was levied on the strength of the beer and on the volume sold. Of most significance was Gladstone's Free Mash Tun Act of 1880. Fundamentally, as tax was to be raised on strength and volume, the brewer was now able to experiment with the raw materials used for brewing. It meant that adjuncts could be used – leading to a broader range of beer types in terms of color and flavor.

Louis Pasteur studied both wine and beer, publishing *Etudes sur le vin* in 1866 and *Etudes sur la bier* in 1876. The year 1876 also saw the founding of the Bacterium Club in Burton-on-Trent, headed up by the three Fellows of the Royal Society based in the major breweries of that small midlands town. The grouping evolved into the Laboratory Club, then the Institute of Brewing, which still exists today as an international organization called the Institute of Brewing and Distilling. It is one of several major societies focusing on the science and practice of brewing, the other notable ones being the

European Brewery Convention, the Master Brewers Association of the Americas, the American Society of Brewing Chemists, and the Brewing Congress of Japan. This technical focus far exceeds anything in the wine industry.

Political and technological advances have continued to impact and shape the brewing industry through the nineteenth, twentieth, and twenty-first centuries. Thus, the crown cork was patented by William Painter in 1892, which together with the developments in glass, enabled the bottle beer market and the possibility of taking the common brewer's beer home as an alternative to drinking it in the pub. Better and better barleys were being bred by the likes of Beaven, Hunter, and Gosset, the last of these being a learned statistician rejoicing in the name "Student" (many will be familiar with Student's t-test). Emil Christian Hansen perfected pure yeast technology in Copenhagen in 1883, a technology (as are so many first evolving in the brewing industry) that is now the very basis of fermentation-based technology from pharmaceuticals to modern biotechnology to, yes, winemaking.

Leopold Nathan developed his stainless steel hygienic cylindro-conical fermentation vessels in Switzerland at the end of the nineteenth century, while techniques for accelerating malting, fermentation, and maturation, as well as maximizing throughputs in the brewery (high-gravity brewing), revolutionized brewing practices without jeopardizing product quality. Brewers were far more prepared to innovate than were the winemakers and were using hygienic, temperature-controlled equipment far earlier than were the vintners.

The first beer was filled into kegs by J.W. Green of Luton in 1946, and the first into steel cans by Kreuger of Newark, New Jersey, in 1935. The first aluminum can was from Coors in 1959. Since then, we have seen better and better plastic bottles and, now, aluminum bottles.

The History of Beer in the United States

The Pilgrim Fathers landed at Plymouth Rock in December 1620 when they would actually have preferred to make landfall further south – why? Because, as the ship's log says: "We could no longer take time for further search or consideration, our victuals being much spent, especially our beer."

There was just enough ale left on board to satisfy the return journey for the sailors. Meanwhile, the less fortunate settlers were left to drink the local water, the bacteria lurking within inevitably making them sick. The boiling stage in brewing (for reasons, of course, unknown at the time) had the great benefit of destroying organisms unwelcome in the gut. Small wonder that the call soon went out for ensuing passenger rosters coming into the colonies to be populated with some brewers.

Adrian Block, from Holland, opened the first commercial brewery in New Amsterdam in 1613. The first paved street in this country was laid in that city in 1657 to aid the smooth passage of horse-drawn beer wagons. Although the early immigrants were, of course, somewhat puritanical, beer was considered then as now to be the drink of moderation. The alternative offerings were suspicious distillates of fermented corn. Indeed, the Scottish and Irish immigrants came with a passion for whiskey, so that just prior to the Civil War, beer represented little more than 10 percent of all the alcohol consumed in this nation. By the eighteenth century, New York and Philadelphia were the principal locales of brewing, and at the turn of the nineteenth century, there were over 150 breweries in the United States producing 230,000 barrels, with one third of the breweries in each of the above two locations.

The early to mid-nineteenth century saw the founding of the great brewing dynasties of the States, all originating in Germany. David Yuengling built his brewery (which still exists in Pennsylvania) in 1829. The Schaefer brothers set up in New York in 1842. In 1840, Lemp in St. Louis and Wagner in Philadelphia opened the first U.S. breweries majoring in lager-style beers. In 1844, Jacob Best founded the company that would become Pabst when his daughter married steamboat captain Frederick Pabst. Bernard Stroh opened his brewery in Detroit in 1850 and, five years later, Frederick Miller purchased Jacob Best's sons' Plank Road Brewery in Milwaukee. In 1860, Eberhard Anheuser purchased a struggling St. Louis brewery and, after his daughter married a supplier named Adolphus Busch, the mighty Anheuser-Busch Company was born. In 1872, another migrant from the Rhineland, Adolph Coors, established his brewery in Golden, Colorado.

73, there were over 4,000 breweries in the United States, outputs some 2,800 barrels each. Rationalization meant that by the conclusion of World War I, there were half as many breweries, each producing more than 50,000 barrels. By the end of World War II, there were only 465 breweries in the United States, averaging some 190,000 barrels.

The production of lager demanded ice for its prolonged storage. Accordingly, such beer had to be brewed in winter for storage (*lagering*) until the greater summer demand. Hence, the merits of locations such as Milwaukee, using the ice from Lake Michigan and local caves for storing the beer. The Wisconsin city rapidly emerged as the great brewing center of the States, with Pabst and Schlitz among those competing with Miller. Once von Linde had demonstrated artificial refrigeration, however, lager could be brewed any time – and anywhere. And the application of Pasteur's proposals for heat-treating beer to kill off spoilage organisms, as well as the advent of bottle and crown cork technology, meant that beer could be packaged for home consumption and consumed almost anyplace. To these innovations can be added the development of the transnational railway network, with railcars containing the latest refrigeration technology, and the advent of cans, with their lighter weight as compared to bottles, and of metal kegs, which allowed for more robust shipping of draught products. All of this allowed the major brewers to ship their products to the great cities across the nation, and the American taste rapidly developed for the pale, brilliantly clear, relatively dry, and delicately flavored products that now represent two-thirds of beer sales in the United States.

The brewers' fortitude was much challenged in the era of Prohibition. Voices against alcohol had been making themselves heard for many years. Dr. Benjamin Rush, a signatory to the Declaration of Independence, argued in 1784 that "ardent spirits" caused *inter alia* obstruction of the liver, jaundice, hoarseness, diabetes, jaundice, gout, epilepsy, madness, and "frequent and disgusting belchings." In particular, he was drawing attention to the impact of the spirits that had been commonplace features of society. A collection of Connecticut businessmen took heed and stopped making rum available to their employees, replacing it with cider and beer, the latter being drinks of moderation. Indeed, President Thomas Jefferson wrote to a friend in 1815 about beer: "I wish to see this beverage become common instead of the whiskey which kills one third of our citizens and ruins their families."

The Union Temperance Society was founded in New York State in 1808. Beer was deemed to be acceptable but the 44 members pledged to "use no rum, gin, whisky, wine or any distilled spirits...except by the advice of a physician, or in case of actual disease, also excepting wine at public dinners." A number of other such societies sprang up, arguing for moderation rather than abstinence.

It was Presbyterian minister Lyman Beecher who became the first to oppose alcohol in all its manifestations. He implored people to join his crusade to rid the country of "rum-selling, tippling folk, infidels and ruffscruff." His sermons were distributed nationwide and, as a result, employers stopped giving drinks to their workforce and liquor rations ceased in the U.S. Army. Beecher's American Temperance Union (ATU) sought to persuade every state to ban the production and sale of alcohol. At first, beer was accepted within the ATU, but that too fell foul of the zealots in 1836. The impact was a *decline* in membership. So many people realized the facts: It was hard spirits that were leading too many astray, not beer.

The fight against alcohol became easier in 1833 when the U.S. Supreme Court ruled that state governments could regulate the liquor trade within their boundaries. Furthermore, it permitted a "local option," in which individual counties and towns could introduce Prohibition if they so wished.

Massachusetts, in 1838, banned sales of spirits in quantities less than fifteen gallons. It didn't last long – customers bought fifteen gallons plus a gill, drank the gill, and then returned the balance. Maine introduced total Prohibition in 1851 and soon thirteen more States had joined Maine, but nine soon repealed the laws or declared them unconstitutional. Only Maine, Kansas, and North Dakota held firm – and in each there were bootleggers and illicit taverns ("blind pigs").

Women soon led the charge against alcohol. One slogan was:

> We do not think we'll ever drink
> Whiskey or gin, brandy or rum
> Or anything that'll make drunk come.

This is hardly classic verse – but at least there was no mention of beer (or wine).

The Women's Christian Temperance Union had prominent members, including the First Lady, Mrs. Rutherford B. Hayes ("Lemonade Lucy"). And they warmly embraced the redoubtable Carry Nation, who declared "hatchetation" in smashing up illicit taverns in her home state of Kansas and beyond, and set off on an enthusiastically received lecture tour in which hatchets could be bought as souvenirs. They do say that no publicity is bad publicity and soon liquor producers were marketing Carry Nation cocktails, and bars were decorated with hatchets and signs that declared "All Nations welcome but Carry."

Carry Nation was probably emotionally troubled for much of her life and therefore the most successful "pro-Prohibition" lobby, the Anti-Saloon League originating in a Congregational Church in Ohio, ignored her. The tactics of this body were more subtle and low key, progressively persuading towns and counties to embrace Prohibition. Soon, they were successful at the state level: Georgia, Oklahoma, and then half a dozen more fell into line. In 1913, after twenty years of existence, the Anti-Saloon League marched on Washington, D.C., with a slogan "A Saloonless Nation in 1920." Several supporters were elected to Congress.

The 65th Congress, convening in March 1917, soon declared war on Germany following the sinking of the Lusitania. This action demanded laws that would ensure that the United States was in a fit state to fight a war, including one concerning the production and distribution of food. A clause was inserted that outlawed the production and sale of alcoholic beverages, such that grain could be conserved. There was disagreement from the opponents of Prohibition, and there was agreement to let the Senate vote on a separate resolution calling for a Prohibition amendment to the Constitution. Astonishing to many, but the Eighteenth Amendment went speedily through Congress and it was ratified by thirty-six state legislatures in little more than a year. Only Rhode Island and Connecticut held out on ratifying the amendment. The amendment was officially adopted on January 16, 1919, with national Prohibition being effected one year later. There was a ban on all intoxicating liquor, defined as a drink containing in excess of 0.5 percent alcohol. Alcohol stocks were destroyed. At least 478 breweries were rendered unable to go about their primary business. One of the biggest names, Lemp in St. Louis, closed its doors forever. Others developed alternative products that their technology might be turned to, such as ice cream, non-alcoholic malt-based beverages (including "near beer"), yeast, and syrups.

The brewers did not "go quietly." Early in 1921, a group of brewers and physicians made efforts to convince Congress that beer was vital medicine. Attention was drawn to its relaxing impacts and to its nutritional merits. It had been suggested that the vitamins in beer had saved the British nation more than once. Attorney General A. Mitchell Palmer opined that doctors could prescribe beer howsoever they believed was appropriate and that druggists could take charge of sales, selling beer from their soda fountains. The Anti-Saloon League was aghast, and Congress quickly limited wine and liquor prescriptions (both of these were deemed "medicinal") to less than half a pint every ten days – and they banned beer (the drink of moderation) absolutely.

It's perhaps not altogether strange that to deny people something that the majority enjoy and don't abuse will inevitably prove unsuccessful. In New York, before Prohibition, there were 15,000 bars. After Prohibition, there were 32,000 speakeasies. Women and youngsters now decided that drinking was something they perhaps should entertain, whereas they hadn't so much before. Booze was coming into the country illicitly from Canada and Mexico, and by ship from Cuba, the West Indies, and Europe. And there was the illegally brewed stuff in the States, much of it dangerous through a lack of regulation. There was plenty of corruption at high levels and of course the making of some infamous criminal reputations among the gangsters, not least Al Capone. Bootleggers collected $2 billion annually, amounting to some 2 percent of the gross national product.

Quickly, there sprang up bodies such as the Moderation League seeking to repeal the Volstead Act, which enabled federal enforcement of the Eighteenth Amendment. In 1930, the American Bar Association adopted a resolution calling for a repeal of Volstead. They were supported by the Women's Organization for National Prohibition Reform. Those advocating "dryness" were at risk of being perceived as defending the gangster culture.

By the early 1930s, the nation was in the midst of the Great Depression. Many argued that it had been brought on by Prohibition and that to repeal the act would help create jobs and put much needed taxation income into the Treasury.

The 1932 presidential campaign was in substantial part fought on the alcohol lobby. Herbert Hoover said that Prohibition had been an "experiment noble in purpose" and he promised to do what he could to correct whatever shortcomings there were. Franklin Delano Roosevelt went a major step further: "I promise you that from this date on the Eighteenth Amendment is doomed."

A campaign slogan was "A New Deal and a Pot of Beer for Everyone." Roosevelt was elected and nine days later he asked Congress to amend the Volstead Act so that the alcohol content of beer could be raised from 0.5 to 3.2 percent by weight. The law was passed. As he sat down to his evening meal on March 12, 1933, Roosevelt is quoted as saying, "I think this would be a good time for a beer." (Not a glass of Chardonnay, I note!).

Whether to enforce Prohibition or not became a state issue – but it took Mississippi until 1966 to emerge from being the last dry territory. For a company to return to brewing after such a hiatus (thirteen years for most states) is no trivial issue. In particular, there had been seepage of trained and skilled brewers and operators and an inheritance of unreliable equipment, leading to equally questionable products in many instances. It was the strong and the resourceful that survived and inevitably this meant strength in size.

By the 1960s, there were fewer than fifty breweries in the United States. But California senator Alan Cranston introduced legislation to legalize home brewing and it was signed into law by President Jimmy Carter in 1978. Of course, there had been plenty of illicit brewing going on before that on a small scale, but the new legislation meant that some of the folks brewing in their garrets and garages could now "come out" and, furthermore, develop their hobby legally and to a much greater extent. In 2005, there were 1,368 breweries in the United States. Most of them (927 to be precise) have a tiny capacity, generally serving beer on the premises (brewpubs). However, plenty of breweries are in the "craft" range of many thousands of barrels. And the movement has spawned enormous interest in a wider range of beer styles. However, we might note that of the approximately 200 million barrels of beer retailed in the United States, only some 7 million barrels is contributed by the so-called "craft" brewing industry, which comprises brewpubs, microbreweries, regional specialty breweries, and contract brewers. In other words, around 85 percent of the beer market in the United States is accounted for by the larger brewers and only a little more than 3 percent by the craft brewer segment. Imports account for the rest.

The Evolution of the Pub

The metal hoops would glisten as they flew through the moist evening gloom, landing with a splat and a "tink!" as they hit simultaneously the target metal pin at the center of the square of soggy clay. Another inch perfect quoit had been thrown in the contest down in the idyllic hamlet of Beckhole in Yorkshire, England. And as the quoits were tossed, so were the team members, officials, and supporters alike sociably and affably quaffing pints of ale, ambers, and browns, and those of a blackness to match the darkening night sky. And this young boy thought that 1950s scene to be fabulous.

It is the ultimate chicken and egg situation. Which came first? The historic game (to read more go to <http://www.tradgames.org.uk/games/Quoits.htm>) or the beer, or rather the pub? Some would have it that quoits were first played by the Romans, but there is no question that it, alongside various other games, became synonymous with the conviviality of the village hostelry. Thus, we have skittles and Aunt Sally, Shove 'alf-penny, dominoes, draughts (checkers), cribbage, and, of course, darts. And what is more quintessentially English than the sound of leather on willow as the cricket game unfolds on the village green. Dispel notions of cucumber sandwiches and tea in china cups. Think mugs of ale.

Surely these images – and their equivalents surrounding the inns, taverns, and other watering holes of other countries – speak powerfully to the central role of beer in the fabric of society. Somehow, wine has not achieved this integral role at the heart of the working person's lifestyle. Its role is far more closely linked to the dining table, a place that beer can with equal justification occupy, in addition to the myriad of other social activities that it lubricates.

We can trace the history of the public house (pub) to Saxon England and the primitive wooden huts (*tabernae*) located at intervals on the roads built by the Romans. A long pole was used to identify such drinking places and if wine was sold as well as mead and ale, then an ivy bush was hung from the pole. By the turn of the second millennium, the alehouse was at the heart of every village. Brewing skills developed in the monasteries were passed to the women brewing in their own homes and they started to sell their beer, essentially converting their homes into what we would now call a "brewpub."

In the early fourteenth century, there was one brewpub for every twelve people in England. In Faversham in 1327, 84 out of 252 traders were brewers. All ale was sold locally because of transport limitations and the difficulty of keeping beer for any length of time. The beer was brewed by females (*brewsters* or *ale-wives*). A *huckster* was a woman who retailed ale purchased from a manufacturing ("common") brewer, while women who sold wine were called *hostesses*. Ale was sold in three types of premises: inns, where you also sought food and accommodation; taverns, which also sold wine; and alehouses. And yet 90 percent of ale was still "home-brew." Alehouses evolved into more up market Public Houses, with different rooms for different classes of customers.

In the "dark satanic mills" of the Industrial Revolution, workers lived in homes that were cozily frugal at best but dark, unhygienic, damp, and cold at worst. For most, it was the pub that offered the warmth, light, and social uplift that they craved. The pub was the heart of the community. It was a mentality that transferred to the United States with the early settlers, who establish the tavern as the heart of the community. It was Dr. Samuel Johnson who said, "There is nothing which has yet been contrived by man by which so much happiness is produced, as by a good tavern or inn." Alas, the pub in the year 2007 is very different. And for that I pin much of the blame on Margaret Thatcher. Let me explain.

In 1989, Thatcher's government introduced the Beer Orders. It was her view that the British brewing industry comprised a monopoly through its vertical integration. There was a "big six" of brewers in the United Kingdom: Bass, Allied, Whitbread, Courage, Watney's, and Scottish and Newcastle. Added to these were countless smaller brewing companies. They all brewed plenty of different beers, but what stuck in Thatcher's craw was the fact they largely sold those beers through their own pubs ("tied houses"). Thus, Bass, for example, had more than 6,000 pubs. The Prime Minister clearly preferred the concept of the "Free House" pubs that were not tied to a single brewer but could sell anybody's beer.

Hence, the Beer Orders were introduced, the first version of which said that no brewer could own more than 2,000 pubs. The potential result was an awful lot of pubs in the market.

However, brewers are nothing if not perceptive. They all knew two things. The first is that the main profit is not to be made in the brewing, but rather in the selling. In other words, the mark-up is in the pub. The second awareness is that if you are going to make a profit in the brewing side and are selling in bulk, then the only way to do that is by brewing lots and lots of beer. And so for all those bigger brewers, the decision was clear: become mega brewers to sell to beer retailers (including the new pub companies that could own as many pubs as they wanted because they didn't brew any beer) or get out of brewing altogether and buy pubs, hotels, and other outlets through which you can sell beer at a healthy profit.

The result of Thatcher's monopolies commission then was that of those Big Six, only Scottish and Newcastle still exist as an independent brewing entity, grown fatter through the acquisition of brewing companies in the likes of France, Finland, Portugal, and Russia. A famous company like Bass, founded in 1777 and with their famed red triangle the oldest registered trademark in the world, is now the world's biggest hotelier (Inter-Continental). Their breweries were divided up between the world's biggest brewing company (Inbev) and Coors.

And so there are indeed still pubs in the United Kingdom but the selection of beers is certainly less than prior to 1989. Not only that, the Beer Orders were a significant factor leading to the demise of traditional styles of English ale. The time-honored route to making ale in the British Isles has been to transfer the fermented beer into barrels, to which is also added a little sugar, some hops, and some isinglass finings. The residual yeast converts the sugar into carbon dioxide, which induces the required small amount of carbonation to the beer. The hops add a pleasing and mellow aromatic character. And the finings clarify the product, allowing the insoluble materials to sink to the bottom of the cask, leaving a clear product. This in itself demands skill on the part of the bar person, for otherwise the customer will get a murky mess in his or her glass. The other point is that the cask, once tapped, has a shelf life of only around three days, for the beer is not pasteurized. The consequence of less than pristine handling of these beers is that the product will become vinegar rather quickly.

In the days of vertical integration, the brewing companies had teams of expert technical staff working with the folks in the pub to ensure that these beers were in tip-top condition. But once this intimate link was broken, the results were often catastrophic.

Easier, then, not to sell these beers at all. Result: the loss of a whole range of beers. Choice and monopoly indeed!

Other forces were, of course, also integral in the changing face of beer drinking in the United Kingdom. We can't blame it all on the Iron Lady. Starting with Guinness, and their widget in the can allowing beer to be dispensed at home in a manner to which it was to be had in the bar, a whole plethora of ales were produced that were designed for consumption away from the pub. It was cheaper drinking – no surcharge slapped on in the bar. And at home there are television and computers and diverse other attractions in an age where it isn't necessary to go to the pub to keep warm and get stimulation.

The pub evolved away from being a place where the focus was chat and simple games to locations for dining (accompanied by wine as often as not!) or for having social interactions of the type not to be had in solitary drinking at home. At the least this would be soccer games on cable and satellite television (cheaper than having your own at home and nearly as good as, and certainly warmer and cheaper, than heading to the stadium itself – where you couldn't drink as freely anyway – Thatcher had interfered in that, too).

Sure, there are still pub games to participate in – but they have evolved in an undesirable direction. Witness, for instance, Beer Pong in the United States: two teams firing table tennis balls at one another's glasses, with the person who receives a ball in their glass obliged to finish the contents of that glass. The difference in pub games from quoits and the like is not a subtle one: In yesteryear, the game was a pastime as the beer was consumed. In latter-day games, the drinking is an integral component of the game. There are no equivalents in the world of wine drinking, an altogether more cerebral activity perhaps. For my part, I would contend that beer drinking should also be cerebral, devoid of stupid drinking games, but not so highbrow as to take beer away from being the true drink of social union.

No matter how you feel about the modern-day pub, it remains at heart a place to drink beer with other people. There simply is not the same inherent community of spirit in a glass of wine as there is in a pint of beer. Beer is real, it simply *is*. There is no need to think or debate, no need to study, romance, or proselytize. Beer speaks for itself. Wine seems to need to have others speak for it – much of it pretentious claptrap.

4. How Wine Is Made

Grapes

The grape is a very sweet commodity, each berry comprising as much as 28 percent sugar. Couple this with the fact that it can possess a very healthy surface population of yeasts only too willing to convert that sugar into alcohol if only they can get at it, then we have ideal conditions for that wonderful, albeit simple, concept known as wine.

The grape vine belongs to the family Vitaceae. The most important member of the clan is the European grape, *Vitis vinifera*, a plant that has been cultivated for more (some would say "much more") than 5,000 years and whose origins are traceable to the region of the Caspian Sea. In turn, there are thousands of cultivated varieties (cultivars) of *Vitis vinifera*, though precious few (perhaps fifty or less) of these have true commercial significance. Some examples, and their characteristics, are listed in Table 4-1. The reader seeking to ponder the pedigree of these various cultivars should go to <www.biology.uch.gr/gvd>.

TABLE 4-1. CHARACTERISTICS OF SOME BETTER KNOWN GRAPE
VARIETIES

Varietal	Characteristics of the grape*
(a) White grapes	
Chardonnay	Apple, pear, peach, apricot, lemon, lime, orange, tangerine, pineapple, banana, mango, guava, kiwi, acacia, hawthorn
Gewürztraminer	Rose petal, gardenia, honeysuckle, lychee, linalool, peach, mango, spice, perfume
Muscat	Terpine, coriander, peach, orange
Pinot Blanc	Almond, apple
Riesling	Woodruff, rose petal, violet, apple, pear, peach, apricot
Sauvignon Blanc	Grass, weeds, lemon-grass, gooseberry, bell pepper, green olive, asparagus, capsicum, grapefruit, lime, melon, mineral, "catbox"
(b) Black grapes	
Cabernet Sauvignon	Black currant, blackberry, black cherry, bell pepper, asparagus (methoxy-pyrazine), green olive, ginger, green peppercorn, pimento
Merlot	Currant, black cherry, plum, violet, rose, caramel, clove, bay leaf, green peppercorn, bell pepper, green olive
Pinot Noir	Cherry, strawberry, raspberry, ripe tomato, violet, rose petal, sassafras, rosemary, cinnamon, caraway, peppermint, rhubarb, beet, oregano, green tomato, green tea, black olive
Syrah	Black currant, blackberry, grass, black pepper, licorice, clove, thyme, bay leaf, sandalwood, cedar
Zinfandel	Jammy (raspberry, blackberry, boysenberry, cranberry, black cherry), briar, licorice, nettle, cinnamon, black pepper

*Note that all of these characteristics need not necessarily feature in the wine, whose character will also derive from processing, including maturation in wood. Source: <http://www.winepros.org/wine101/grape_profiles/varietals.htm> on July 26, 2006.

Other species of Vitis have some importance. *V. lubruscana* is a grape that is native to America (it formerly grew wild in the mid-Atlantic states). It is often called the "fox grape" on account of its production of the chemical methyl anthranilate, which possesses a distinctly foxy character. The "muscadine" grapes are of *V. rotundifolia*, with their strong sultana-like character.

American varieties are rather more robust than are their European cousins, and so there has long been a desire to cross them. Vines tend to have bud-containing shoots (the scion) from *V. vinifera* grafted onto rootstocks of other *Vitis* species. The scion and rootstock must be compatible with one another and with the local soil and climate.

Most wineries grow their own grapes; however, they may also purchase them from nearby vineyards, allowing a degree of financial flexibility and the wherewithal to take advantage of periods of economic advantage, for example, bumper crops. Many criteria are used when a winemaker chooses the cultivar to be grown, including yield, flavor, time of ripening, resistance to pests, and the best fit to the local climate and soil type. All of the environmental impacts that are brought to bear on vineyards are encapsulated nowadays by the word *terroir*.

The grape cultivar must perforce be matched to a growth location. This cannot be too equatorial because conditions will be too humid, leading to a risk of disease. In equatorial regions as well, there is not the period of inherent dormancy that is necessary to ensure that the grapes develop to best advantage at the prescribed time. Neither should the region be too polar, for then winters are excessively cold and the growing season too short to allow sugar build up. In practice, the best grapes for wine are grown within the latitudes of 50° and 35° on either side of the equator.

Grapes do not fare well in any locale that is excessively hot, cold, or windy. This means that the favored vineyards are those on gentle slopes and in valleys that are sheltered from anything other than the mildest breeze. Having said this, it is said that grapes giving the most interesting wines are from vines on the cooler fringes of the climatic bands.

It seems that soil is less significant than weather in the cultivation of wine grapes. Indeed, less fertile, stony soils may sometimes be preferable. It is important that the soil be neither too acidic nor too alkaline. If fertilizer needs to be added to boost the nitrogen level in the soil, it must not be to excess; otherwise, there is wasteful growth of plant material without increased grape yield. There is also an increased risk of spoilage and of developing the toxic substance ethyl carbamate in the ensuing wine.

The roots of vines are extensive and tolerant of drought. This is good, because excess moisture presents a disease threat. The ideal is warm, dry summers, no frosts, and an efficient irrigation system. The soil should drain well. Some regions have endemic diseases, such as Pierce's Disease, and phylloxera (an insect that attacks rootstock). Pierce's Disease is due to a bacterium called *Xylella fastidiosa* that is spread by an insect known as the glassy-winged sharpshooter (*Homalodisca coagulate*). The problem is restricted to regions with mild winters. The multiplying bacteria block the water-conducting system in the vines and therefore cause dehydration and wilting.

Excessive summer rain, or too much irrigation, increases the risk of powdery or downy mildew. Furthermore, if the berries take up too much water, they swell and burst, allowing mold growth. Some form of pesticide treatment is frequently necessary in regions particularly susceptible to disease and infestation.

Vines break out of dormancy in the spring when the average daily temperature rises above 10°C (50°F). As the vine grows, the berries increase in size about five-fold, going from small green pea-like entities that lack sugar and possess high levels of acid to the sugar-loaded berries needed by the winemaker. These sugars comprise an equal mix of glucose and fructose, arising through the hydrolysis of sucrose. As well as these sugars, the flavor components characteristic of the various cultivars also develop during growth, as do the anthocyanins that afford the hue to the red varieties

The rate at which the vine grows and with which the grapes develop is very dependent on the local environment. To illustrate using California's central valley, the vines start to sprout early in April and the grapes are ready for picking ("the crush") in mid-September. In cooler regions, the planting of early-ripening varieties is advisable.

A grape grown in a short season will contain less sugar and so tend to produce lower alcohol, tart (high-acidity) wines, whereas warm climates allowing sustained growth will militate for the production of high-sugar grapes appropriate for the production of dessert wines.

And so wines from geographically remote regions that enjoy a common climate have clear similarities. These likenesses may be disguised if the winemaker employs techniques such as sugar addition (chaptalization) and acid reduction (the malo lactic fermentation; see later). By using such practices, the winemaker may overcome seasonal differences. For as many winemakers as are prepared to do this, there are those who rejoice in the vagaries of nature, ergo we have the concept of vintage.

The pruning of the vines is a skilled task. If the pruner is too heavy-handed, then an excessive number of buds (ergo berries) will be sacrificed. If the pruner errs excessively on caution and removes too few buds, then grape quality will suffer. There will usually be 500 to 600 vines per acre. Vines are trained up stakes in the first growing season, a single shoot alone being allowed to develop on each stake. Pruning of vines occurs in winter months when the vines are dormant and the canes have hardened.

Microorganisms

The sugar fungus *Saccharomyces cerevisiae* thrives naturally on the grape, which means that there is, strictly speaking, no need to seed a wine fermentation with yeast. These days, the majority of winemakers do add yeast so as to achieve greater consistency and more predictability in their wine. Even so, many winemakers would opine that the yeast makes relatively little difference to wine flavor, and is merely pertinent for the amount of alcohol that it will generate, at what speed and with regard to how easily it can be removed prior to bottling.

Other Saccharomyces species may have a role in the fermentation of some wines, notably *S. bayanus*. Additionally, strains from genera such as Pichia, Torulaspora, and Kloeckera are relevant in some wines, but they are far more sensitive than is Saccharomyces to sulfur dioxide and they do not relish alcohol concentrations exceeding 4–6 percent. They can only have an impact before the alcohol content has been raised above this level by Saccharomyces and before any sulfiting is performed.

The flor yeast, *S. bayanus*, is used in the production of some sherries. It produces quantities of acetaldehyde ("green apple") by growing on the surface of a fermentation batch and consuming oxygen.

After sugar (which provides carbon and energy), quantitatively the next most important ingredient demanded by yeast is a source of nitrogen. Although the grape does contain amino acids that can be used by yeast, levels can sometimes be insufficient, and the wine-maker may add diammonium phosphate to ensure that the yeast is not starved.

Grape juice is somewhat acidic, with a pH of 3–4 (neutrality is 7 on a scale of 1–14). At this pH, many other organisms fare badly and so this helps select for Saccharomyces to perform to advantage. Mold growth is blocked by the carbon dioxide that is generated in fermentation and the alcohol too is anti-microbial. Just as for beer, no pathogens or toxin-producing organisms can thrive in wine.

The lactic acid bacteria are important in the production of many wines. They perform the malo lactic fermentation, in which the highly acidic malic acid is turned into the substantially less acidic lactic acid. As there is a direct link between acidity and sourness, this means that wines subjected to this step are less tart. It is an important process stage for wines derived from grapes grown in relatively cool environments but is undesirable in wines from grapes grown in warmer regions, as it lessens the "bite" deemed important to flavor balance. Malo lactic fermentation may be used for the top-of-the-range full-bodied reds but far less so for fruitier whites.

Another organism that can have a major impact on wine is *Botrytis cinerea* ("grey mold"), which attacks unripened grapes that encounter excessive humidity, for example, where summer rains are prevalent. The impact is a cracking of the skins, which allows other microbes to adversely affect the yield of fruit and its quality. On the other hand, limited growth of the mold prior to a return to low humidity conditions allows some loss of moisture through the punctured skin, a shriveling of the grape, and a consequential strengthening of sugars and flavors. The mold also removes some acidity and delivers its own interesting flavors. In this situation, winemakers speak of "noble rot" and rejoice in the conversion of these grapes into very sweet, rich wines, including the French Sauternes and the Hungarian Tokays.

Some Key Components of Must and Wine

The sugar concentration of a must is quantified in terms of specific gravity, and declared as either degrees Balling or Brix. As well as for monitoring fermentation (as the sugars are converted into alcohol, which has a specific gravity less than 1, the Balling/Brix values fall), these values are used to identify the best time at which to harvest. By multiplying the Brix value by 0.55, one can approximate the alcohol content that will be achieved in a wine.

Sulfur dioxide (SO_2) is a fundamental component of most wines, and is usually present at levels between 15 and 25 mg/L. SO_2 fulfils three major functions: It inhibits undesirable microorganisms, it prevents browning of white wines, and it protects wine from oxidative deterioration. Oxidation may also be restricted by other additions. Although banned in some countries – for example, the United States – potassium ferrocyanide is used to remove excess iron from some wines; a material that promotes the oxidation process. Surely it is ironic that this is a treatment tolerated for wine in Germany, a country possessed of some highly restrictive, almost puritan attitudes when it comes to beer (under the terms of the Reinheitsgebot).

The Crush

As someone so long involved in the brewing industry, where beer is produced (to use the common shorthand) 24/7, it always amuses me that all hell lets loose in a winery for, what, three months at most? I am ever curious about what those guys do the rest of the year (although it has been rumored that they drink a significant quantity of beer). The answer, of course, is that there are far fewer people the rest of the year, the crush employing a sizeable temporary workforce. Harvesting of grapes usually occurs in August through September and October. The time of harvesting helps determine the sweetness/acid balance of grapes. Grapes grown in warm climates tend to lose their acidity more rapidly than do those from cooler places, due to the removal of malic acid as the grapes mature.

Ripe fruit is crushed immediately after picking. However, if white grapes are picked in hot conditions it is best that they are chilled to less than 20°C (68°F) before crushing. Indeed, many will pick them by night, which is not really feasible with manual labor but is with machinery sporting headlights. There is no such problem for red wine grapes, which are fermented at higher temperatures.

Fruit destined for white table wine is picked when its sugar content is between 23 and 26° Brix. However, grapes going to red table wine have a longer hang time.

Payment is based on the Brix value. The winemaker will also insist on other specifications, including the amount of non-grape material that can be tolerated and that the berries be free from mold.

Grape Processing

One metric ton of grapes yields roughly 140 to 160 gallons of wine. The basic stages involved in passing from one to the other are relatively few: The grapes are crushed to generate the must, which is fermented, and then there is some degree of downstream processing prior to packaging. As we shall see in the next chapter, the production of beer is very much more complicated.

All vessels in the wine-making process these days tend to be stainless steel and jacketed so that the temperature can be regulated. Furthermore, there is an increased appreciation of the importance of maintaining hygiene, so modern tanks are fitted with so-called "in-place cleaning systems" (CIP). Tanks and pipes between batches can be cleaned successively with water, caustic soda, sequestering agents (which remove built up salts), and a sterilant, such as the dilute bleach one might use about the home. CIP is far more extensively used in brewing.

Grapes are moved by screw conveyors from a receiving tank to the stemmer-crusher. The must passes from there either to a drainer, a holding tank or (in the case or red grapes, where the skins are not removed before the yeast is added) directly to the fermenter.

The stems are usually removed from crushed grapes so as to avoid off-flavors. However, Pinot Noir is sometimes fermented in the presence of stems in order to yield its characteristic peppery aroma.

In stemmer-crushers, the grapes are smashed using either spinning blades or squeezed through rollers. In both cases, the crushing is into a perforated drum that separates grape from stem.

Soft or shriveled grapes are harder to break open. The amount of breakage needs to be carefully controlled. Damage must not be excessive, otherwise undesirable materials will be extracted from the grape cells, skin, and seeds.

It is not necessary to immediately separate juice from skins for red wine, but is so for white or blush wines. The materials that afford the color to wine are located in the skin – these are molecules called polyphenols, indeed, a particular type called anthocyanins. Blush wines are lighter than rosé wines. For rosés, an overnight contact of juice and skin with limited fermentation allows the appropriate level of extraction of color components. After the juice destined for rosé or blush wines has been separated from the skins, it must be protected from oxidation by the addition of SO_2.

SO_2 may already have been added in the crushing stage. How much SO_2 is added to the crusher is influenced by several factors. If there is a risk from mold in the vessels or if there is a propensity for air pick up, then the SO_2 addition may be significant. If the grapes aren't infected and the opportunity for air ingress is low, then SO_2 may even be avoided, but then the juice should be kept cool ($<12°C$, $<54°F$).

The speedy separation of skin and juice for white wines also lessens the opportunity for astringent tannins to accumulate in the wine. However, it also means that some desirable attributes may not fully be realized, for example, the flavors that give the pleasing sultana-like character to Muscat wines. Clearly, there is a skilled balance to be struck regarding oxygen availability, SO_2 use, contact time, and temperature.

"Thermovinification" may be used for some cheaper wines to enhance color recovery. This involves the rapid heating and cooling of crushed grapes. Heating kills the cells, which therefore release pigments. However, potentially undesirable flavors are also released.

Once the grape structure has been disrupted, there is initially a large volume of liquid released which is able to be drained off – this is called "free run." By pressing, more juice is extracted, so-called "press run." The residue is known as pomace.

There are various designs of press with differing degrees of severity. Membrane or bag presses are very gentle and leave little sediment. Bladder presses are often used for rapid processing, but the juice tends to contain higher levels of solids.

To accelerate juice settling for a clearer product, pectic enzyme is often added in crushing. Pectin originates in the grape cell walls and is a very viscous and sticky molecule (jam makers are well aware of its merits in solidifying preserves, but the winemaker wants high liquidity, not a solid lump.) By adding pectinase, which disrupts the pectin, pressing is easier, juice flows more freely, and the yield is higher.

Fermentation

As red grapes are fermented with their skins in place and losses during pressing are much less significant, the yield is about 20 percent higher than for whites. Modern fermenters are likely to be constructed from cleanable stainless steel. However, wooden vessels would allow some access of oxygen and this generally permitted some of the desirable flavor changes to occur.

After the juice has been separated from skins (if it is), it is held overnight in a closed vessel. Then it is racked or centrifuged off the sediment before the yeast is added. On the whole, winemakers seem to be less fastidious about their yeast than are the brewers. Increasingly, they will employ widely available dried yeast, rather than a "house" yeast of the type championed in the production of beer. The yeast needs some oxygen in order to make its cellular structures, and the necessary aeration usually occurs *after* the introduction of yeast, to avoid the development of color due to the action of enzymes that meld oxygen into the tannic components of the must.

White, rosé, and blush wines are fermented at 10–15°C (50–59°F), reds at 20–30°C (68–86°F). Of course, the yeast performs more rapidly at the higher temperatures, and under these conditions tends to generate substantially higher proportions of flavor substances such as the fruity esters. The aroma of white wine is very dependent on the extent to which the fruity esters are produced during fermentation.

As the alcohol concentration rises, so is there interference with the action of the yeast, especially at higher temperatures. Accordingly, fermentation gets progressively slower. High levels of undesirables, such as the eggy hydrogen sulfide, can arise if fermentations are sluggish from the start. The character derived from different grape varietals is better preserved at lower fermentation temperatures.

Color and flavor extraction from red grapes is maximized by mixing the contents of the fermenter through pumping or by rousing. Twice a day, around half of the vessel contents will be pumped over. The extraction of flavor and color is greater at higher temperatures and as the ethanol concentration increases through fermentation.

In situations where grapes don't ripen well because of a short, cool growing season, the winemaker may add sugar (sucrose), to a maximum of 23.5 Brix. However, it is illegal to do this in some locales, such as California. Colder climates also make for an unfavorable acid balance in grapes. The sourer malic acid is predominant in grapes from cooler environs, whereas the mellow tartaric acid predominates in grapes grown in warmer places.

The pH during fermentation should be kept below 3.8, allowing wines to ferment more evenly, avoiding malo lactic fermentation and letting better sensory properties develop. Also SO_2 is more effective at lower pH. This pH control is especially important for white wines, in which the pH may be lowered to 3.25–3.35 by the addition of tartaric acid.

Fermentation should be complete within twenty to thirty days. Wine is usually racked off the yeast ("lees") at the end of fermentation. Some winemakers leave the wine in contact with the yeast for several months (*sur lie*), perhaps with occasional rousing, to promote release of materials from yeast that benefit flavor.

In different regions of the world, some unique enological practices may be conducted. One example obtains for the production of Beaujolais wines, a technique known as *Maceration carbonique*. It affords wines with distinct estery (pear, banana) characteristics. Whole clusters of grapes are stored in an atmosphere of CO_2, allowing the limited conversion of sugar to ethanol and the production of several flavorsome substances. The berries are held at quite a high temperature of 30–$32\,°C$ (86–$90\,°F$). The weight of the grapes and the action of the developing ethanol and carbon dioxide cause the cells within the grape to break down, releasing their sugary, flavorsome, and colorful contents. After a week or a week-and-a-half, the grapes are pressed and the juice obtained combined with that previously collected by free running. Then, the fermentation is allowed to proceed to dryness at 18–$20\,°C$ (64–$68\,°F$) before SO_2 is added and the wine clarified.

Aging

Wines, more so reds than whites, are felt to benefit from aging, either in tank, barrel, or bottle. Some Chardonnays are aged in oak barrels, which provide some flavors. Burgundy and Loire whites are left *sur lies* for up to two years.

Red wines are aged after malo lactic fermentation. Bordeaux wines are held two years in barrel. By comparison, the aging of Zinfandel should not be prolonged so as to retain the raspberry character.

Clarification

White wines are clarified either by centrifugation or by chilling and treatment with bentonite (a clay) or silica gels, derived from sand.

Both of these agents adsorb protein that might otherwise precipitate in the bottled wine and ruin its appearance. Cold treatment also removes tartrates that might otherwise deposit in the wine. Some winemakers add the milk protein casein or an agent called polyvinylpolypyrrolidone (PVPP) to adsorb polyphenols, which are also implicated in turbidity formation by sticking onto proteins in the self-same reactions that are employed in the tanning of leather. Red wines are usually fined to reduce their astringency: Proteins such as gelatin, egg white, or isinglass will bind those tannic components that afford most astringency.

The filtration of wine is relatively uncommon and only applied when really necessary, such as to recover wine from the lees, after cold stabilization treatments, or immediately before bottling. Some winemakers will filter their wine through membranes to remove traces of microbial infection.

Just like beer, wine is susceptible to the threat of oxygen. In particular, enologists worry about color development ("pinking") and may add ascorbic acid to help SO_2 counter this. Oxidation is promoted by traces of metal ions such as iron, and these may be removed by adding casein or citrate to the wine. Another addition made by some winemakers is copper, to remove traces of hydrogen sulfide that will give an odor of rotten eggs to the wine.

Champagne/Sparkling Wine

The best champagnes are produced from the juice of Pinot Noir or Chardonnay grapes. There must be scrupulous avoidance of color development, and so the production of champagne involves the extensive use of SO_2, bentonite, and PVPP.

The carbon dioxide is made by the action in the bottle of strains of S. *bayanus* that are flocculent and tolerant of high alcohol concentrations. The yeast, parent wine, and invert sugar (sucrose – the type of sugar you and I put into our coffee but which has been split into its constituent parts glucose and fructose) are put into pressure-resistant bottles that are then sealed with a crown cork of the type used in bottling beer. A small headspace will be left in the bottle before it is laid on its side and held at $12°C$ ($54°F$). The wine ferments to dryness over several weeks but may be left in excess of a year for the best quality. There follows the stage of "riddling," wherein yeast is worked into the neck of the bottle. The yeast is loosened by hitting the bottom of the bottle with a rubber mallet or by using a shaking device and then the bottle is put into a rack such that it is held neck down at an angle of 45 degrees. The bottles are rotated a quarter turn daily until the yeast sediment has all arrived at the cap. Then the bottles are chilled to $0°C$ ($32°F$) and taken through a brine bath that is cold enough to yield a frozen plug of wine. The crown cork is removed and the ice plug allowed to slide out, taking the yeast with it. Then the bottle is immediately turned upright, refilled with more wine containing sugar and some SO_2, and corked.

For sparkling wines, very cold riddled wine is completely removed from bottles, pooled, and cold-stabilized under pressure. Then it is filtered and returned to bottles for corking. Some wines are carbonated by bubbling with carbon dioxide prior to bottling, just as is the case for many beers.

Any residual oxygen may be removed by sparging the wine with nitrogen gas and thereafter air ingress must be carefully avoided. Some winemakers add sorbic acid to sweet table wines to counter the risk of spoilage by bacteria.

5. How Beer Is Made

THE BREWING OF BEER IS SUBSTANTIALLY MORE COMPLICATED than is the making of wine. The brewing scientist in me says that it is not too challenging a task to tread a few grapes, and then leave them alone while the native yeast does its thing and converts the grape sugars into wine. A bit of cleaning up here and there and then, voila, wine. I am, of course, being utterly cynical, and retain my admiration for the skill of the winemaker in taking the right grape from the right locale and turning it into a delightful product.

Yet, I still insist that the complexity and skill of the succession of folks who grow barley and hops and then turn them into beer is perforce greater: The whole journey toward a bottle of beer is far more exacting than is the making of wine. There are many more steps. And most brewers insist that the product meet stringent specifications, both in terms of flavor profile and in the levels of a diversity of analytical measurements. Some of the molecules that contribute to flavor are specified to parts per billion levels. The barley and hops are no less prone to seasonal variation than is the grape. But brewers overcome these fluctuations so as to achieve a consistent beverage, whereas the winemaker will tolerate them and the attendant variation (and sometimes, unpredictability) in the finished drink.

It's exactly analogous to the operation of a Boeing 777. There is every control and safeguard imaginable, but the skill comes in navigating the prevailing conditions. Brewers seek the still air and avoid the turbulence. Winemakers plow straight into it – and charge the customer a premium for what might sometimes be a decidedly bumpy ride. The alcohol in wine is derived from the sugars derived from fruit, for the most part the grape, but we all remember Gramma's prune wine and Uncle Bert's elderberry champagne.

For beer, the sugars are derived from grain. The predominant cereal for making beer worldwide is barley, and has been since time immemorial, as we saw in Chapter 3. In part, this is because the hull that is retained on barley when it is threshed has traditionally acted as a filter bed for separating the liquid extract from the solid residue in the brew house. However, barley malt itself (as we shall see) affords specific characteristics to beer, rather different ones from other cereals. And so other beers have their own styles and natures because they are made from other cereals. There are some great wheat beers in the world, products that tend to be somewhat cloudier than barley-based beers, and fruitier, spicier, and fizzier. In Africa, especially, there are beers made from sorghum and they are altogether different again, indeed, not to the taste of many Westerners at all. Oats, rye, corn, rice, millet – the list goes on – they are all used in the production of beers.

All of these cereal grains are replete in starch. It is the food reserve that provides succor to the embryonic plant residing within each and every one of those kernels. Of more relevance to the brewer is that this starch will yield the sugars that he or she will provide to the yeast to convert into alcohol. But, unlike in the case of the grape, these sugars are not readily available. The grape is basically a bag of sugars, ready for action. Break it open and, off you go, the yeast has its meal. Starch, by contrast, is altogether too big a challenge for yeast: We might say that it is over-faced. To be accurate, yeast can't digest it. Rather, the starch, which is a polymerized form of glucose, must be broken down to its constituent parts (or, rather, for the most part, to pairs of glucoses held together as a couplet called maltose) before it is in a form that delights Saccharomyces. The journey from starch wrapped up in solid grain to maltose in a liquid form ready to be enjoyed by yeast is a long and convoluted one and takes weeks, not the hours involved in dishing up a meal of grape sugars for the wine yeast. Here we will take that journey for barley alone.

Barley

Just as there are many cultivars of grape, so too are there many varieties of barley, either six-row (*Hordeum vulgare*) or two-row (*H. distichon*); depending on how many rows of grain there are on the ear. Only some of these varieties are suitable for making into beer. In fact, the bulk of the barley crop worldwide (more than 130 million acres) is grown for feed purposes, either to sustain animals or to convert into foodstuffs for the human. Every cook knows of the merits of pearl barley. The minor proportion is malting barley, so named because through controlled germination it can be converted into "malt," which is the staple ingredient of beers as well as other foodstuffs, ranging from baby food to breakfast cereals, and whiskey to soup. Only the malting varieties will germinate easily and in a way that when the malt is subsequently dealt with in the brewery, it will readily give high yields of fermentable sugars.

Barley can be divided into winter and spring types. The hardy winter varieties are sown in the late fall and remain dormant over the winter months before getting a head start in the growth stakes over the spring varieties, which are sown in March or April. As a result, yields tend to be greater from the winter cultivars and thus may be harvested a little earlier in the subsequent fall. Some brewers insist that the beer made from those barleys is inferior – pretty much a dogmatic belief not founded on any substance, but no less opinionated a stance than we see for those adherents to certain wine cultivars.

Barley is a major crop in temperate and tropical regions, better capable of withstanding heat and salinity than other crops, such as wheat. A total of almost 150 million tons is produced annually, in more than 100 countries, with Russia, Canada, Germany, and France topping the list. As there is more space for the individual grains to develop in a two-row cultivar, they accumulate more starch and, as more starch means potentially more sugar for the brewer, they are favored.

Unlike grapes (and hops, as we shall see) barley is a relatively tolerant plant. This is not to say, however, that the husbandry involved in growing the best barley for malting is trivial. One requirement is that the amount of protein allowed to accumulate in the grain should be limited. For a kernel of a given size, the more protein is present the less space there is for starch. Thus, one of the stipulations for the growing of malting barley is that the use of nitrogenous fertilizers is strictly limited. This means, in turn, that the yield of plants is reduced. A farmer has to be persuaded to grow malting barley – and, as is inevitably the case whenever persuasion is needed, the incentive is money. The malting premium is the extra value that a malting grade crop will attract over a feed-grade crop. Woe betides, however, if there is any deficiency in the crop. If rejected for any reason, then the farmer will at best receive only the feed rate. Small wonder that many farmers prefer to avoid the risk, and therefore grow the higher yielding, fertilizer-tolerant feed barleys or alternative cereals.

Barley is far from being immune to all manner of threats, whether fungal or animal. Diseases such as mildew are more likely in wetter regions, as is the threat of infection by an organism such as Fusarium, which can introduce a small protein that causes the beer to spontaneously surge out of the bottle or can and deposit itself on your pants. It's a heinous problem called gushing and mightily inconvenient for brewer and customer alike.

The climate and terrain have an enormous impact on the properties of barley. The ideal is a cool, damp growing season that allows sustained growth succeeded by warm and dry weather when the grain is ripening and ready for harvest. Failing this, there are risks of poorly filled grain (low starch), high levels of other carbohydrates that cause problems for the brewer in processing, and the introduction of dormant status into the grain. Dormancy is a natural condition that prevents the seed from germinating prematurely on the ear of the plant. However, if it prevails in the harvested grain it is a problem for the maltster, because it means that the grain has to be stored for the dormant condition to become naturally alleviated (as it will) – but this takes time and means that the grain cannot be converted into malt at the time the maltster would prefer.

One other key criterion for malting barley is that it should be alive. Dead grain cannot be malted. So, the farmer must be very careful not to kill the corns. In wetter regions, it may be necessary to dry the grain so that it can be stored without risk of infection. Such drying is a very delicate operation that must be conducted to perfection so that the embryos are not destroyed. All in all, then, we see that the concept of *terroir* fairly applies to barley the very equal to grape.

Malting

No brewer can make excellent beer by using barley directly. This is because the grain contains very low levels of the enzymes that are needed to convert the starch into a fermentable state. Not only that, barley possesses a fairly astringent, drying character: Try chewing some barley grains and you will find that they will irritate at the back of the throat, quickly dry out the palate and, if you try to bite them, their toughness will present a clear dental risk.

These are the reasons why the grain is malted. Barley is first steeped in water, a process that takes approximately two days. When the grain takes up water, it springs into action and makes the enzymes that are needed to break down the food reserves that it contains. Some of these enzymes soften the structure of the grain, rendering it more readily milled. Also produced are the enzymes that will break down starch. They don't have much of a chance to act in the malting operation, but will come into their own later in the brewery. While the grain is being softened, there is the commensurate production of nutrients that pass to the baby plant to nourish it. This is evidenced by the appearance of rootlets.

After a period of controlled germination, which may be as long as a week or so, the grain is dried in a stage that lasts one to two days. This halts the germination – otherwise, there will be wasteful growth of embryo and a diminution of the precious starch – but also there is a driving off of the undesirable raw flavors in the "green" malt (notes reminiscent of bean sprouts and cucumber, nice in their own right but not as a component of beer) while introducing the pleasing malty flavors. Try malted milk chocolate balls or a bedtime malted milk drink such as Horlicks to get a feel for malty character. At the same time, the heating leads to a development of color in the grain. The more the malt is heated, the darker the color. This is the reason why ales tend to be darker than lagers: Historically, the grain is heated to a higher temperature. Not only does this mean more color but it also means that there is a different flavor spectrum. Ales tend to be maltier, with more toffee-like characteristics, while lagers tend to be lighter in flavor as well as paler in color. In both ale and lager malts, the heating has to be conducted in a very restrained manner, with the temperature being raised gradually, for otherwise there will be a destruction of the starch-degrading enzymes that have not yet fulfilled their mission.

For the so-called specialty malts, the heating is more prolonged and intense. Progressive heating leads to more intense toffee characters, and when the temperature gets very high in special units called roasters, there is the development of very dark colors but also the burnt, coffee-like, mocha, chocolate, and acrid characters that, in balance, give the hue and the flavor to products like porters and stouts. Table 5-1 gives a listing of malts and adjuncts used in the production of beer. These are the materials that are used as additions alongside the predominant pale malt to afford the rich diversity and color to beers – myths about the abundant use of molasses and caramels need to be put to bed.

TABLE 5-1. MALTS AND ADJUNCTS

Product	Details	Purpose/comments
Pilsner malt	Well-modified malt, gentle kilning regime not rising above ca. 85°C	Mainstream malt for pale lager beers
Vienna malt	Similar to Pilsner malt but higher nitrogen (N), more modification, final kilning temperature ca. 90°C	Mainstream malt for darker lagers
Munich malts	From higher protein barleys (e.g., 1.85% N), prolonged germination, low temperature (e.g., 35°C) onset to kilning to allow stewing (ongoing modification), then rising temperature regime to curing at over 100°C	For darker lager beers
Pale malt	Relatively low N (e.g., <1.65% N), well-modified, kilning starting at ca. 60°C and rising to a final curing temperature ca. 105°C	Mainstream malt for pale ales
Chit malt	Very short germination time and lightly kilned	Permissible as adjuncts in countries such as Germany with restrictions such as Reinheitsgebot
Green and lightly-kilned malts	No or restricted kilning after substantial germination	Alternate to exogenous enzymes
Diastatic malts	High N barley (especially 6-row), steeped to high moisture content, long cool germination, gibberellic acid if permitted, very light kilning	High enzyme potential for use in mashing with high levels of adjuncts

(*continued*)

TABLE 5-1 *(continued)*

Product	Details	Purpose/comments
Smoked malts	Kilning over peat	For beers with smoky character, e.g., Rauchbier
Wheat malts	Germinated wheat, usually somewhat under modified, lightly kilned (e.g., < 40°C)	For wheat beers
Rye malts		For specialty beers
Oat malts		For specialty beers, including stouts
Sorghum malts	Steeps may incorporate anti-microbials such as caustic; warm germination (25°C)	For sorghum beers; malted millet may also be used as a richer source of enzymes
Cara Pils (a caramel malt)	The surface moisture is dried off at 50°C before stewing more than 40 minutes with the temperature increased to 100°C, followed by curing at 100–120°C for less than 1h	To afford color and malty and sweet characters to lighter beers
Amber malt	Pale malt is heated in an increasing temperature regime over the range 49–170°C	To afford bread crust, nutty characters to beer and color
Crystal malt	As for Cara Pils, but first curing is at 135°C for less than 2 hours	To afford toffee, caramel characters to beer, and color
Chocolate malt	Lager malt is roasted, by taking temperature from 75–150°C over 1 hour, before allowing temperature to rise to 220°C	To afford chocolate, roast, coffee, burnt, bitter characters to beer, plus color
Black malt	Similar to chocolate malt, but the roasting is even more intense	To afford harsh, astringent, roast, burnt notes to beer, plus color

TABLE 5-1 *(continued)*

Product	Details	Purpose/comments
Roasted barley		To afford sharp, dry, burnt, and acidic notes to darker beers (n.b., drier than roasted malts)
Raw barley		Added in mashing as a cheaper source of extract
Torrefied barley	Barley heated to 220–260°C	Easier to mill than raw barley and starch is pre-gelatinized
Flaked barley	Grain rolled immediately after torrefaction	Does not need to be milled
Raw wheat		Adjunct for wheat-based beers
Torrefied wheat	Wheat heated to 220–260°C	Easier to mill than raw wheat and starch is pre-gelatinized; wheat-based adjuncts may be used for barley malt beers to enhance foam
Flaked wheat	Grain rolled immediately after torrefaction	Does not need to be milled
Wheat flour	Fine fraction produced in milling of wheat	Mash tun adjunct not requiring milling
Corn grits	Produced by milling of de-germed corn (maize)	Add to cereal cooker for gelatinization; sometimes for economic reasons, or for production of lighter flavored and colored beers
Corn flakes	From torrefaction and rolling of corn	Does not need to be milled or cooked in brew house
Rice grits	Produced by milling of de-germed rice	Add to cereal cooker for gelatinization; for production of lighter flavored and colored beers
Rice flakes	From torrefaction and rolling of rice	Does not need to be milled or cooked in brew house

(continued)

TABLE 5-1 *(continued)*

Product	Details	Purpose/comments
Cane sugar	Refined from sugar cane	Sucrose – for addition to kettle as wort extender or beer as priming agent
Invert sugar	Cane sugar after hydrolysis to fructose and glucose	For addition to kettle as wort extender or beer as priming agent
Corn sugars	Produced from the hydrolysis of corn starch by acid and/or enzymes	Range of products for addition to kettle depending on extent of hydrolysis. At one extreme is high dextrose sugar (approaches 100% glucose) and at the other extreme is high dextrin syrup. Latter for body – very low fermentability. Former for high fermentability (e.g., in production of light beers). Most widely used is high maltose syrup – sugar spectrum reminiscent of that from conventionally mashed malt

Source: Table reproduced from Lewis, M.J. and Bamforth, C.W. (2006). *Essays in Brewing Science*, Springer, New York.

In just the same way that it is possible to articulate the flavor characteristics of grapes (see Table 4-1 in the previous chapter), so can we speak of taste and aroma notes in malts. A glance at Table 5-2 of this chapter, however, will reveal a very different set of descriptors. Whereas, quite understandably, grape flavor terminology is suffused with fruity and spicy terms, for malt we have words that very much speak to cereal and grassy characteristics, and certainly for the

TABLE 5-2. FLAVOR NOTES ASSOCIATED WITH MALTS

Cereal	Cookie, biscuit, hay, muesli, pastry
Sweet	Honey
Burnt	Toast, roast
Nutty (green)	Bean sprout, cauliflower, grassy, green pea, seaweed
Nutty (roast)	Chestnut, peanut, walnut, Brazil nut
Sulfury	Cooked vegetable, dimethyl sulfide
Harsh	Acidic, sour, sharp
Toffee	Vanilla
Caramel	Cream soda
Coffee	Espresso
Chocolate	Dark chocolate
Treacle	Treacle toffee
Smoky	Bonfire, wood fire, peaty
Phenolic	Spicy, medicinal, herbal
Fruity	Fruit jam; banana, citrus, fruitcake
Bitter	Quinine
Astringent	Mouth puckering
Other	Cardboard, earthy, damp paper

Terminology developed by Brewing Research International.

darker malts, to the roasting event. Much more main course than dessert, one might say?

In the Brew House

The malt houses tend to be located close to the malting grade barley. So in North America, that means they are for the most part in U.S. states such as North Dakota, South Dakota, and Minnesota, and in the Canadian provinces of Alberta, Manitoba, and Saskatchewan. Once produced, the malt has to be stored, usually for at least a month, before it can be used in the brewery. Nobody as yet knows why, but it is certainly the case that freshly kilned malt does not "perform well" in the brew house. Again – it's not like the grape – tread and go!

The grain will be shipped to the brewery by truck or train. Breweries tend not to be located in such picturesque environs as the wineries and are for the most part in regions of heavy industry. That is, after all, historically where the thirstiest people have been.

The malt is first milled into much smaller particles, to a size that will be readily hydrated by the water added in a proportion typically of three parts water to one part milled grist. The water may be at a temperature as low as 45°C (113°F) in order that some of the breakdown processes started in the maltings may be completed. However, all mashes will be taken, either immediately or after twenty minutes or so, to 65°C (149°F). The brewing vessels have jackets through which steam can be passed in order to regulate temperature, with 65°C as a critical compromise temperature. It is here that the starch, which is in the form of tough little balls ("granules") in the grain, is melted in a process called gelatinization. Some people call it "pasting" as it is literally akin to what happens when you make a gluey paste in the kitchen. Most importantly, the starch in this pasted form is now much more susceptible to attack by the starch-degrading enzymes and, within an hour or so of this mashing, the starch is digested into a mixture of sugars and so-called dextrins, which are small chains of incompletely degraded starch. These latter products cannot be handled by yeast, but altogether about 80 percent of the starch is converted into a fermentable form.

The liquid, called wort, is now drained through the residual grain material, which is mostly hull. The remnant solids are trucked off as cattle feed. These grains will rapidly deteriorate if not moved

off-site quickly and, in any event, they can't be allowed to accumulate; otherwise, there will be no space for new brews in the vessels.
The logistics are finely tuned – not a trivial matter.

The wort is boiled for about one hour with hops in a vessel called
a kettle, the purpose of which is to stop enzyme activity, steril-
ize the wort, precipitate unwanted material that might otherwise
emerge as a haze in the finished beer, and partially concentrate it
to extract the valuable components from the hops. It is also at this
stage that unwanted "raw" aroma compounds are driven off from
the liquid. Indeed, it is this stage that you can smell when you pass a
brewery.

Water

One of the biggest differences between beer and wine is that to make
the former a huge amount of water is used, whereas for the latter,
there is need for relatively little. In making wine, there is a crushing
of the grapes but little or no addition of water unless there is a need
to thin out the must. In other words, pretty much all of the water
in a bottle of wine originates in the grapes. Such is not the case for
beer. Not only are huge amounts of water needed in the malting process, but typically three times as much water as grain is used in the
mashing stage, with yet more used to sparge the grain in lautering,
quite apart from the large volumes dedicated to keeping the plant
in pristine cleanliness and also that is converted into the steam used
to transfer heat in the brewery. In the best running breweries, there
may be a demand for perhaps four to five times more water than ends
up in the finished beer. In the least efficient breweries, there may be
ten times more, or greater.

All of this water needs to be of top quality. It needs to not only be free from infection and any flavor taint or color (and most brewers will pass the incoming water through a filter to ensure its quality), but it also needs to have precisely the right salt content. The extent of hardness (levels of calcium and magnesium sulfates, chlorides, and bicarbonates) has a profound effect on what happens in the brewing process. If those salt levels are too low, then salts will be added. If too high, they can be removed in special filtration techniques. A brewer can match any water in the world in this way. And so a brewer in Germany wanting to mimic the famed ales of Burton-on-Trent, England, where the water is phenomenally hard, can add calcium sulfate. In fact, they give a good old German word to the process ("das Burtonization").

Surely, when folks use the word terroir – and when we think of natural resources and mother earth – we should have cascading through our mind's eye pure sparkling water with those salts derived from rocks. Isn't it, then, a word more apt for use with beer?

Hops

If malt is the soul of beer, then hops are the spice. How unusual to have a plant that really has only one major use. Although there is an increasing likelihood of more and more pharmaceutical applications for the hop, the only sustained application of hops is in the production of beer. The Romans used them in salads, but that I wouldn't recommend. I once tried aftershave made of hops but my wife threw it away after one splash. It is said that you dream of the person you love if you sleep on a hop pillow, but it has never worked for me. And I don't recommend smoking them, despite the fact that their closest familial relative is cannabis. It has been said that hops aid those with poor digestion and who have problems with their intestines. Hops are said to work wonders for constipation, overcome premature ejaculation, and ease anxiety.

Hops (*Humulus lupulus*) grow between the latitudes 30° and 50° either side of the equator – in North America, the key states are Washington, Oregon, and Idaho. These latitudes are pretty much where you find that the best wine grapes are grown. The similarities do not end there.

The hop plant is a hardy perennial. Its rootstock remains in the ground year on year and is trained annually onto a trellis framework, with the plants growing as high as sixteen feet, though some of the newly bred dwarf varieties grow half as high and are all the more easily handled for that very reason. Soils need to drain well, but irrigation is also very important. Hops represent an extremely delicate crop, one that is susceptible to all manner of pests and diseases, such as *Verticillium* wilt, downy mildew, mold, and the damson-hop aphid, or hop fly. Varieties differ in their susceptibility to infestation and have been progressively selected on this basis. Nonetheless, it is frequently necessary to apply pesticides, which are always stringently evaluated for their influence on hop quality, for any effect they may have on the brewing process and, of course, for their safety.

Hops are dioecious, in other words there are separate male and female plants. It is solely the female plant that is of direct value to the brewer, because it is the flowers of the female, the hop cones, that contain the valuable ingredients. As hops can be grown from cuttings then the male is shoved aside, particularly because his presence leads to a fertilizing of the females and the production of seeds that most brewers think is a bad deal for beer.

In the hop cones are multitudes of tiny yellow lupulin glands that contain resins and oils. The resins are converted in the boiling stage of brewing to a more soluble form that is the source of the bitterness in beer. The more hops added and the longer the boil, the more bitter will be the beer. The oils afford the aroma that is so characteristic of many beers. Like other essential oils, these are quite volatile and so if all the hops are added at the start of boiling, the aroma is entirely driven off through the flue and the beer will have no hoppy character. In the production of lager beers, therefore, perhaps 5 to 10 percent of the hops are not added until quite late in the boiling stage, which allows some of the oil to be extracted and retained. In the production of ales, the traditional practice has been to add the vast majority of hops at the start of the boil, but to hold back just a handful of the hops to add to each filled barrel. This gives the more complex "dry" hop character of ales, altogether less subtle than the "late" hop aroma in lagers.

The bitter compounds have significance far beyond taste alone. Their presence is a major reason why beer displays stable foams. They are antiseptic, and are one of the causative agents in making beer one of the most resistant of foods to microbial infection. The downside is that they are sensitive to light – and if exposed to sunlight or even artificial illumination they degrade to give a compound with the decidedly unglamorous aroma of skunk. This is the reason why some brewers don't use hop cones in the kettle, but rather extract the resins from hops with cold, liquid carbon dioxide. These brewers then chemically change the extracted resins into a form that is no longer sensitive to light and that can be added to beers destined for clear glass bottles, with no fear of skunking. Other hop products include the pellets, in which the hop cones are milled and compressed into a form that disintegrates during boiling giving better "utilization" of the key components.

TABLE 5-3. AROMA DESCRIPTORS FOR HOP VARIETIES

Hop	Aroma
Bramling Cross	Distinctive strong spicy/black currant
Brewers Gold	Black currant, fruity, spicy
Cascade	Flowery, citrus, grapefruit
Chinook	Spicy, piney, grapefruit
Fuggle	Delicate, minty, grassy, slightly floral
Hallertau	Mild and pleasant
Hersbrucker	Mild to semi-strong, pleasant, and hoppy
Millenium	Mild, herbal
Saaz	Very mild with pleasant earthy hoppy notes
Styrian Golding	Delicate, slightly spicy
Tettnang	Mild and pleasant, slightly spicy
Willamette	Mild and pleasant, slightly spicy; delicate estery/black currant/herbal

Source: <http://www.winepros.org/wine101/grape_profiles/varietals.htm>   and <http://www.wellhopped.co.uk/Variety.asp> on July 26, 2006.

Just like grapes and barley, there are numerous varieties of hops that differ in their levels and composition of resins and oils. Some varieties contain high levels of resins and therefore are chosen by those seeking high bitterness levels with less consideration of aroma. Other varieties have wonderful oil compositions, highly prized for the aroma that they provide to beers. Just as for grapes, the interaction between variety and environment is very important. *Terroir, une autre fois.* Table 5-3 lists the characteristics of some hop varieties. The reader might recognize similarities with Table 4-1 for grapes in the previous chapter, but there is far less description, nay affectation. Beers might fairly be marketed on the basis of hop "varietal" just as much as is a wine on a grape varietal. Invariably, they aren't. Perhaps this whole terminology should be beefed up.

There is absolutely no reason why it couldn't be. However, brewers are inherently fair-minded. The truth is that many of the characteristics of hops are modified during brewing, fermentation, and maturation, so that it is not necessarily an easy matter to taste a beer and say, "Ah! Fuggles" or "Unmistakably Hersbrucker."

Fermentation

The boiled wort is clarified by filtering through residual hops or, if pellets or hop extracts are used, by swirling in a large vessel called a whirlpool, with all the precipitated material collecting in a pile in the middle of the base of the vessel. (It was a circumstance first witnessed by Albert Einstein when stirring his tea – he saw the leaves accumulating in a pile in the center of the bottom of the cup.) Then the wort is cooled and injected with air or oxygen because the yeast needs some oxygen to satisfy the production of some of its cellular parts. Otherwise, the wort, with its sugars to satisfy the need for carbon and energy and its amino acids to supply nitrogen and various minerals (notably calcium and zinc), is pretty much a stand alone food source. Sometimes a little extra zinc needs to be added.

Just like the winemakers, brewers measure the strength of their feedstock in terms of specific gravity. Rather than speaking of Brix, however, brewers use either the specific gravity *per se* (the weight of the wort as compared to the weight of the same volume of water) or use degrees Plato, which is basically the percent sugar in the wort, comparable to Brix. And just like the winemaker, the brewer monitors the course of the fermentation by following the decrease in the specific gravity/Plato as sugar is converted into alcohol.

Unlike most winemakers, brewers are fastidious about their yeast. Most brewers have their own yeast strains, perhaps different ones for different products. They know with confidence how those yeasts will behave in the brewery, at what rate they will convert the sugar into alcohol, to what extent they will stick together ("flocculate"), whether they will sink or remain in suspension, and the extent to which they will make certain flavor substances.

The ale strains belong to *Saccharomyces cerevisiae*, the lager ones being *S. pastorianus*, which we believe is a hybrid strain arising many millennia ago from the merging of *S. cerevisiae* with *S. bayanus*, strains of which we encountered in Chapter 4 in the context of champagne production and flor. For both ale and lager yeasts, there are many varieties, and brewers are convinced that their strain in each case is the right one for their beer. In truth, the blander the beer, the more important is the strain of yeast. If you have a robust stout, with its strong roast malt and bitter character, the subtleties of flavor introduced by the yeast are largely inconsequential. However, for a beer with little malty or hoppy character, then the fruity, estery character introduced by the yeasts is much more significant.

Another important factor in fermentation is temperature. Ales will tend to be fermented at a higher temperature than will lagers, perhaps 15–25°C (59–77°F) as opposed to 6–12°C (43–54°F). This means that ale fermentations are much faster, with the yeast belting out more fruity characters in a manner comparable to red wine fermentations. Lagers, by contrast, tend to be characterized by more sulfury notes, directly as a consequence of the slower, colder fermentations.

As well as alcohol, carbon dioxide, and flavors, the other main product of brewery fermentations is yeast, about three times more than was "pitched" in at the start. Some of this can be used to seed the next fermentation providing it has all the characteristics deemed important, most important of which is that it should be alive (viable). In practice, as long as the alcohol concentration produced in the previous fermentation is not too high (say, less than 6 percent) this is likely to be the case. However, if a yeast has labored to produce a very strong beer, then it is fit only for disposal. This is the reason why producers of stronger alcoholic beverages, such as wine, use the yeast only once.

Most brewers will only use one batch of yeast for five or six successive fermentations, after which newly propagated yeast is used. The old yeast makes delightful food for pigs – which especially enjoy the residual alcohol, so they are relaxed and merry as they pile up the pounds. In the United Kingdom, the yeast is converted to an extract for spreading on toast known as Marmite; in Australia, it's Vegemite. You either love it or hate it.

Maturation

Brewers differ considerably in the extent to which they keep beer in tanks after fermentation (lagering) and before filtration and packaging. In all instances, they ensure that beer is left in contact with yeast long enough for the latter to remove the last traces of two substances that give undesirable characteristics to beer. The first is diacetyl, which affords a butterscotch flavor, all right in Chardonnay (and popcorn) but certainly not in the majority of beers. The second substance is acetaldehyde, which has an aroma of green apples or emulsion paint and is death to any beer, even if it is a feature of certain sherries.

There may be other subtle nuances of flavor that convince some brewers that prolonged aging is warranted, but frankly, many of the arguments are unproven, more founded on marketing strategy than technological realism.

Downstream

Brewers are fastidious about avoiding air ingress downstream. The other thing they have to do is remove materials that might lead to haze development in the final product in trade. To counter this, they chill the beer to as cold a temperature as possible, say -1 to $-2°C$ (28–30°F), to precipitate and settle out solids. Then they remove still soluble haze-sensitive proteins using silica hydrogels, papain, tannic acid, or soluble polyphenols through the addition of PVPP. The beer may be cleared of already insoluble material by the action of isinglass, which is derived from the swim bladders of certain fish. Then the beer is filtered, using either diatomaceous earth (kieselguhr) or volcanic ash (perlite) as filter aid. The carbon dioxide level will be adjusted to the level demanded for the beer-style in question and then the product is either canned, bottled, or kegged.

The Packaging of Beer

The simple truth is that the most expensive component of a bottle of beer is the bottle, especially if you factor in the tremendously expensive bottling operation. To enter into a bottling hall is to encounter high tech. Conveyors transport bottles through washing cycles to the fillers, which are huge spinning tanks like over-inflated whirring doughnuts. Each of these contains the beer and numerous filling heads. These heads contain tubes that are lowered with precision into the bottles as they line up and are raised into position on the swirling behemoth. Through the tube a vacuum is drawn in the bottle and then it is flooded with carbon dioxide to ensure no trace of air. A valve opens automatically to allow beer to gently flow into the bottle and the carousel keeps turning, more and more bottles continually joining such that the entire circumference of the filler is adorned with bottles at various stages in the filling cycle. When the bottles are full, the tube slides upwards again and the bottle strikes gently against a "tapper," which causes bubbles to form and foam to rise from the surface of the beer, sweeping away air from the neck. This air, if left in the beer, would cause it to stale too quickly. The moment that the liquid beer starts to seep from the top of the bottle, the crown cork is pressed into place. All of this occurs at speeds that might be more than 1,500 bottles every minute. Filling cans is a very similar process, save that it may proceed at more than 2,000 cans every minute. These are mighty machines, far more dynamic, intense, and precise than those found in the typical winery.

There has long been interest in the packaging of beer in plastic bottles, producing a light-weight product and enabling the opening of markets in which glass is not desirable or possible, for instance, within the sports stadium. Whether plastic bottles are any more aesthetically satisfying than cans is a moot point, but in any event, the barrier to acceptance is literally that: the barrier. The plastic originally used to encase beer was excessively permeable to air, meaning that the beer staled very quickly. The amount of air ingress through latter-day plastics is much less, but there are still those who resist the concept of beer in plastic, rather like wine aficionados resist screw top bottles.

Beer for the taps in the tavern is filled into kegs, aluminum or stainless barrels that go to and from the brewery to carry beer to the thirstiest customers. In each is a spear, which is multipurpose. It is through this that the kegs (usually inverted) are washed – by coupling a pipe to the head of the spear and blasting in water, then caustic, then more water, and finally sterilizing with the hottest steam. Then the keg passes to another head that carefully floods the vessel with carbon dioxide to blow away air, and then fills it with beer. Thence to the warehouse and on to trucks for shipping to the cellars of bars, where a different coupling will be made to the head of the spear, this one for the purpose of serving the product. When the bar person pulls the handle, carbon dioxide from a tank in the cellar is forced into the beer and, because the liquid has to go somewhere to make space, it leaves through the dispense pipes and appears through the tap and makes its way into your glass.

There is a strong lobby of support in countries such as England and Wales for "cask-conditioned ales." These are not matured in the brewery, but instead at the end of fermentation when, still containing residual yeast, they are filled into barrels. These casks, traditionally of wood but these days usually fabricated from the same materials as kegs, feature two holes: one close to the base of one of the narrow ends and the other in the center at the widest part on one side of the cask. Each hole is plugged with wood. The first of these holes is where the spout is hammered in when the cask is "tapped." The other hole (the "spile") is where the beer is going to "breathe."

Alongside the beer at fill, the brewer also adds a handful of hops (to introduce the pronounced dry hop character characteristic of these beers), some priming sugar (for the residual yeast to use), and isinglass finings (hydrolyzed collagen protein derived from the swim bladders of fish caught in the South China Sea). The barrel is rolled and then shipped to the pub. The spile is filled with a hard wood piece that is relatively impermeable to gas, notably the carbon dioxide that is produced by the yeast using the priming sugar to naturally carbonate the beer. Once the carbonation level is reached, the spile may be replaced by a somewhat more permeable piece. Meanwhile, the yeast and other insoluble material in the beer are settled out by interaction with the isinglass finings.

The result is a low carbonation beer, with more "hoppiness" and tremendous drinkability. It is classically drunk at cellar temperatures (15°C/59°F), inaccurately referred to by Americans as "warm," although relative to the 4°C/39°F at which the average Californian takes his or her beer (i.e., refrigerator temperature), I guess it is. The downside to these beers is that they are prone to contamination by acetic acid bacteria – remember, they are not sterilized and are basically open to the air, so vinegar production will occur without the care and attention of a conscientious cellar man.

Control and Automation

The production of cask-conditioned ale is a very time-honored process. It speaks to an age of art and craft. By contrast, modern breweries are bastions of close control. The brewery control center resembles NASA headquarters. The facility is fitted from start to finish with in-line sensors that can monitor all the key pulses of the process and relay the information back to the computer screen in "space HQ." Thus, the most fabulous control is achieved and tremendous consistency is possible in the products.

6. The Quality of Wine

I N CONSIDERING THE QUALITY OF ANY ALCOHOLIC BEVERAGE, it is insufficient to discuss just the obvious characteristics of a product, which in the case of wine would be color, clarity, aroma, and taste. Perhaps more so for wine than any other beverage, including beer, quality is shrouded in a mystique, a culture, a state of mind, and an almost intangible meeting of geology, climate, locale, tradition, art, mystery, and hyperbole.

Let me illustrate. A year or two back, an entrepreneur by the name of Fred Franzia decided that he was going to shake up the snobby world of California wine with the release of products marketed under the name Charles Shaw, but colloquially known to all as "Two Buck Chuck." This was on account of the fact that they retailed in the charming chain stores of Trader Joe's at the princely sum of $1.99. Understandably, these wines were very popular with all – everyone, that is, except those who considered themselves wine aficionados, experts, and commentators, who decried the exercise as a loss-leading gimmick. That was until the day that the toffee-nosed judges selected Charles Shaw 2002 California Shiraz as one of 53 finalists for the International Eastern Wine Competition at Corning, New York, in a blind tasting of 2,300 submissions. The judges were seen next morning at breakfast to stagger around, heads held low, utterly ashamed of their ignorant mistake and completely baffled as to how they could have been "misled." Perhaps it did not occur to them that actually the wine was pretty darned good. Rather that they would take the message that, perhaps, just perhaps, there is no simple correlation between cost and quality. There is no question that this wine would not have come within a million miles of being a finalist if the wines had been identified.

This competition was certainly not the first time that so-called wine connoisseurs were flummoxed. Most famous of all was the Paris wine tasting of 1976. Parisian merchant Steven Spurrier organized it, and nine experts were judging "blind," including the supposed top eight wine experts in France. They started with the whites, and the French vintages were blown out of the water by California Chardonnays. It was the same story with the reds, with a Cabernet Sauvignon from California being the top wine in the whole competition. One taster said of one wine that it spoke of "the magnificence of France." It was actually a Cabernet from the Napa Valley. "That is definitely a Californian wine – no nose" was said about a Batard-Montrachet '73.

Connoisseurs? Maybe they were. All the exercise actually proved was that there are some outstanding wines from the New World, not all of them fetching stupid prices. Which speaks entirely to my point: The quality is written as much in the label, the brand, the vintage, and the chateau as it is in the molecules that make up the liquid.

Exactly as is the case for beers, there seems to be a drinkers' learning curve with wines. Just as the youthful imbiber invariably selects low bitterness beers, the young wine drinker prefers sweet lighter whites, and will progressively move to more robust, aged, and complex reds.

Leaving to one side the impact of bottle, label, and ritual, then we might consider wine quality according to the criteria of appearance, smell, taste, and touch. The last of these of course refers to "mouthfeel" or "body."

Whether we discuss wine or beer, a major judgment of anticipated quality, including flavor, will be determined by the simple act of looking at the product in a glass. Happily and sensibly for wine, the simple act of transferring the drink into a separate receptacle for imbibition is very much the norm. Alas, for beer, it seems increasingly to be an exception, meaning that product appearance is altogether less relevant. Brewers have much to do to dissuade the customer from this vulgar practice: How much more valuable a product might they have if they built up the theater of the pour to be at a level with wine (avoiding, we hope, the frankly rather silly "nosing" of glass and cork before the waiter is invited to dish the stuff out).

One appearance feature of wine that is seldom associated with beer is "legs" or "tears" or "church windows." These are the streams that flow down the inside of the glass after a wine has been swirled. The physics is complex, the direct relevance to quality is questionable, but the visual impact is pleasing.

The majority of wines are clear – that is, they are free from turbidity or sediment. Invariably, precipitates will develop in aged wines – think of vintage ports – in which case the wines should be decanted, taking care not to disturb the precipitate.

Wines are fundamentally of three colors – red, pink (viz. rosé, blush, White Zinfandel) and pale yellow (a.k.a. white) (Table 6-1). There may be subtle or not so subtle variations of hue, but the breadth of appearance hardly compares with beer, where the range extends from colorless to black.

TABLE 6-1. THE COLOR OF WINE

Wine	Color
(a) *White table*	
White Riesling	Greenish yellow to yellow
Chablis	Light yellow
Chardonnay	Yellow to light gold
Sauvignon blanc	Yellow to light gold
(b) *Red table*	
Pinot Noir	Low to medium red
Cabernet Sauvignon	Medium red
Zinfandel	Medium red
(c) *Sweet table*	
Sauternes	Light gold to gold
Tokay	Light gold to amber gold
Rosé	Pink
(d) *Sherry*	
Fino	Light amber yellow
Oloroso	Brownish gold to amber
Dry baked	Light amber
Sweet baked	Medium amber
(e) *Red dessert*	
Tawny port	Amber-red
Ruby port	Ruby red
(f) *White dessert*	
Muscatel	Light amber-gold to gold
White port or Angelica	Medium yellow
Madeira	Medium amber
Marsala	Dark amber
(g) *Vermouth*	
Dry	Light yellow
Sweet	Medium to dark amber
(h) *Champagne*	
Red	Red
Rosé	Pink
White	Light yellow

Source: Amerine, M. A. and Singleton, V. L. (1977). *Wine*. University of California Press, Berkeley.

The aroma and taste of wine should match the expectation delivered by appearance. Unquestionably, wine flavor is complex and is proudly championed by wine aficionados who draw attention to this in portraying the mystique and complexity of the beverage. Indeed, it is claimed that there are some 400 separate compounds that contribute to the aroma of wine. (The reader may, however, like to ponder on the fact that there are probably twice as many relevant substances in beer.)

The flavors of wine come from grape, by yeast metabolism in fermentation, and during aging. It must be borne in mind that some of the characteristics developed during fermentation gradually decay during aging. Some of these changes may be desirable, for instance, the gradual purging of compounds containing the sulfur atom, substances that tend to have aromas variously described as eggy, vegetable, or cheesy.

It is far more common for wines than beers to be described with affected, abstruse, and prosaic descriptors. We read of hillsides and scents of herbs and delicate little numbers and fresh spring mornings, and often it is hard to relate this to some of the frankly dodgy "plonk" that sits in the glass in front of us.

Nonetheless some of the language is rather blunter as a meaningful guide to how a wine actually tastes. In just the same way that "wet dog" has been employed as an honest descriptor for a character that can develop in beer, so is it sometimes said that a certain wine best left nameless has the nose of a gooseberry bush newly urinated upon by a cat.

The basic *taste* in both wine and beer, as in any other foodstuff, is encompassed by the sensations of sweet, sour, salt, and bitter, the characteristics determined by interaction of substances with the taste receptors (buds). However, most of the flavor is determined as *aroma*. This may be direct – by sniffing the headspace above the wine as one would when approving the product presented by the somme-lier. Or it may be indirect – the so-called retro-nasal effect – when the aroma compounds are released in the mouth and pass into the nose through the passage at the back of the oral cavity.

Two words have come into play in the world of wine to describe odor. The first is aroma, which some restrict to the character deliv-ered by the grape, and the second *bouquet*, which speaks of the char-acter developed in the vinification process and storage. Semantics, surely?

Quantitatively, the most important component of a wine is alco-hol, with its warming impact but also sweet sensation and impact on *body*. Glycerol, too, contributes to sweetness and body.

A selection of acids is present in wine, such as tartaric, malic, and citric from the grape and lactic and succinic contributed by yeast during fermentation. Wines deficient in these substances have a flat-ness, whereas excessive levels of course make the product sour. At a low pH (high acidity), red color is enhanced and there is less of a tendency for wine to "brown" through oxidation. Furthermore, the sulfur dioxide becomes more effective. Of course, some of the acids contribute some specific aroma characteristics – for example, acetic giving vinegary and butyric buttery.

The body of wine and its astringency are separate but related phe-nomena, owing much to the contribution of phenolic and polyphe-nolic constituents. The mouthfeel or body is also determined by its alcohol content and perhaps by diacetyl, a substance produced by yeast and which has the character of butterscotch.

The polyphenols are generally of much more significance for wines than they are for beer. The science of this fascinating class of substances is extremely complicated. The so-called flavonoids contribute to color, taste, and mouthfeel. They can be divided into three fractions: anthocyanins, flavanols, and flavans.

The anthocyanins are located in the skin of the grape and are responsible for color. Their role in the grape is to attract birds.

The flavanols, such as quercetin, are color precursors. Thus, they are found in the skin and to a much lesser extent in the pulp. They may contribute to bitterness in red wine and to co-pigmentation by interacting with anthocyanins. Their role in the plant is to protect the grape from ultraviolet light.

The flavans are responsible for bitterness and astringency, thereby deterring predators. They also participate in co-pigmentation. The flavans are located in both the grape skin and in the seeds. Catechin and quercetin are major types, as are the proanthocyanidins, such as leucocyanidin, which oxidize to produce anthocyanins. As these molecules polymerize, there is a shift from a bitter impact to one of astringency.

The levels of these various components found in a wine are very dependent on processing conditions, such as extent of contact with skin and grapes, and the degree of pumping over. Thus, there tend to be higher levels of flavans in red wines, with the darker color being an obvious manifestation of this factor.

The flavor components of wine are sometimes divided into *primary* aromas, which are directly derived from the grape; *secondary* aromas, which develop during vinification, notably in fermentation; and the *tertiary* aromas, which arise in storage.

There is some knowledge of specific chemical compounds that are derived from grape varietals and that afford specific character to the wines made from them. Thus, methyl anthranilate ("foxy") contributes to the aroma of Lambrusca, 2-methoxy-3-isobutylpyrazine ("bell pepper") to Cabernet Sauvignon, and damascenone ("rose") to Chardonnay. Muscats contain terpenes such as linalool and geraniol (compounds also found in hops). Gewürztraminer contains 4-vinylguaiacol, the self-same molecule that puts the clove-like spiciness into wheat beers. Sauvignon Blanc has a guava-like character arising from 4-methyl-4-mercaptopentan-2-one. There is rose-like 2-phenylethanol from Muscadine grapes.

Some of these flavor active compounds are bound up with sugars in the grape. Yeast makes enzymes that sever the link between the flavor-active molecule and the sugar over time, and thus there is a time-dependent release of flavor during fermentation.

Other flavor substances are developed by the yeast through its metabolism. These include the esters, which give fruity characters. There are aldehydes, most notably acetaldehyde, which is a major contributor to the aroma of Fino sherries but an excess of which will make a table wine taste "flat." There are alcohols other than ethanol and also amines and sulfur-containing compounds. Of course, because Saccharomyces (albeit different strains) is the organism involved in producing wine *and* beer, there is no added complexity pertaining to wine in this respect. For both types of drink, the important variables in fermentation that will impact how much of the various compounds are produced include the strength of the sugar solution (the stronger it is, the higher the level produced of many of these substances), the temperature of fermentation, the amount of oxygen the yeast gets, the hydrostatic pressure, and the yeast strain itself. In the last of these, winemakers are relatively unadventurous.

The maturation of wine impacts on aroma but also on taste attributes such as astringency. There are fundamentally two different types of maturation. That occurring in vat or barrel involves extraction of substances from the walls of the barrel and also changes induced by oxygen, such as the polymerization of polyphenolics, leading to a softening of astringency because of the precipitation of the very large molecules produced. There may also be a decrease in acidity. Changes in bottle do not involve extraction or oxidation, but reflect the interactions between various constituents of the wine.

Most notable among the taints in wine is trichloroanisole, which causes "corked" musty or moldy flavor. The logical way to guarantee avoiding it is to use plastic corks or screw caps. This is beyond the aesthetic pale for many, but this conservative snootiness is changing. Taints may also arise from wooden vessels and substances responsible may include geosmin, which gives an earthy note and 2-methylisoborneol, a medicinal aroma. They arise due to the chlorine treatment of corks and the subsequent action of bacteria and molds. Corks should be maintained at very low moisture content to minimize this problem.

The shape of the drinking glass seems to impact the perception of flavor. It is said that daffodil-shaped glasses will cause wine to be directed at the palate in regions that are alternatively rich or denuded of certain taste buds, for example, those which detect tannins that lead to astringency.

Finished wine is rather resistant to spoilage. The very high alcohol content, low pH, and presence of sulfite all contribute, but many hold that it is the polyphenols that are the most important antimicrobials. However, it is possible to have mold growth, albeit it might take a microscope to see it. A visual manifestation of it might be gushing of the product.

7. The Quality of Beer

T HE QUALITY OF BEER, IF WE WERE TO DISTILL IT INTO A nutshell, may be described as "all as for wine, plus bubbles." Certainly there are wines that effervesce, but in no instance do the bubbles survive as stable foam. Despite the fact that too many people in the United States chug their beer directly from can or bottle and a visitor to London will encounter ale (albeit delicious) that resembles cold tea, every image of beer that you will encounter on screen or in print features a rich, dense, creamy foam.

There is no question that foam impacts drinkers' perception of a product. Show customers images of beer with good or poor foams and those with superior foam are declared to be better brewed, fresher, and better tasting. All this is based solely on appearance; not a drop has been drunk in making this evaluation.

So why do most beers display stable foams whereas other drinks, such as champagne and sodas, do not, despite the fact that often times they contain more carbon dioxide than does beer? The answer lies in the presence in beer of molecules that stabilize the bubbles: They are carried into the head when the latter is created and, once in the bubble walls, they fashion a framework (liken it to scaffolding) that prevents the foam from collapsing. The major substances involved are classes of proteins that originate in malted barley (and in adjuncts such as wheat, but not in corn or rice), and the bitter substances that derive from hops.

This being the case, it is generally true that the more bitter the beer, the more stable its foam. Likewise, the higher the proportion of malt and especially of wheat, which has especially powerful foaming proteins, the more stable the foam.

It is not simply a matter of whether the foam is retained. Foam quality can be described according to several criteria. First and foremost, how much foam is formed? The more carbon dioxide (CO_2) in the beer, the more foam is produced. So it is harder to produce a head on a traditional English ale, which might contain only around 1.2 volumes of CO_2 (one volume would be one milliliter of gas for every milliliter of beer), than it is on a Bavarian wheat beer, which might contain significantly more than 3 volumes. In all cases, work has to be exerted to generate the foam. That might be a beer engine (the traditional hand pump) for cask ale. It might be a vigorous pour by hand from a bottle – too often, I have been to restaurants where the server attempts to pour the beer ever so gently down the side of the glass in order to *prevent* foam formation and get as much liquid as possible into the glass. Excuse me! Pour the beer directly to the center of the glass at the bottom, allow the foam to form with gusto, and then leave me the bottle to top things off at my leisure and for others in the restaurant to admire my brand selection.

Surely, the most contemptible way of causing foam to form is the widget, lumps of plastic or metal in cans and sometimes bottles, which serve to "nucleate" the foam, so that when you open the container and start to pour the gas breaks out with vigor. It is important to ensure that the beer is chilled before attempting to open the container; otherwise, the beer will spray out with gay abandon. This is so despite the fact that the very beers that contain widgets – stouts and ales – should not generally be consumed at refrigerator temperatures.

More elegant, and much more simple, is to pour beer into scratched glasses – simple scoring of the base of the glass. Bubbles form more readily at these scratches, and streams of them will ascend through the glass – so-called "beading" – to form and continually replenish foam.

Next, the foam must remain, which is where those proteins and bitter acids come into their own. The choice of bittering agent is relevant. Some brewers use so-called "reduced" bitter acids. These have been modified so that they no longer degenerate in the presence of light to form the reprehensible skunky character that is associated with beers in green or clear glass bottles that do not contain these altered molecules. The reduced bitter acids have particularly strong foaming properties and give heads that are thick and stiff, sometimes to an excessive degree.

The bitter acids are responsible for converting foams from liquid to quasi-solid in nature. This takes a little time – so when next you pour your beer allow it to stand for a minute or two and you will see the change in texture. And it is this transition that makes for lacing or cling: The solid-like foam will tend to stick to the side of the glass to form a pattern that is felt to be attractive by many people – but not all, some women in particular thinking it looks unattractive and dirty. In fact, it is very much a sign of a clean glass.

Indeed, there are far more problems with beer foam due to dirty glasses than with any shortcoming in the beer itself. Fats, such as are introduced from food, and detergents, such as residual dish cleaning liquid, kill foam. Best not to have greasy lips, or moustaches to which globules of grease can adhere, or lipstick if you seek to avoid collapsing the head on your beer. And never, ever, wear a mustache and lipstick. When washing glasses, the trick is to use good soapy water to wash away any oily residues, but then to rinse the glasses thoroughly in clean water to eliminate detergent, before allowing the glasses to drain naturally: Don't wipe them on a grimy cloth. And certainly don't carry glasses by plunking your sticky fingers inside them.

Foaming can be excessive of course – to the point of spontaneity. If a container is opened and the contents immediately shoot out, then we have a gushing problem. It may be a consequence of injudicious handling of the container – it has been shaken – in which case, allowing it to stand for a few hours should rectify the problem. However, the problem may be more insidious than that and could be a result of the presence in the beer of other substances. If that is the case, no amount of storage of the beer will solve the problem. The most frequent cause of gushing is a small protein made by a fungus called Fusarium that can infect grain. Good agricultural practices should obviate the threat.

Most beers are "bright" and free from haze and certainly from sediments. This is not unequivocally the case. In particular, many wheat beers are turbid due to the presence of yeast and other insolubles. These are the Hefeweissens, as opposed to the Kristalweissens, which are clarified.

Many substances can contribute to insolubilization reactions in beer. Most notable is a bridging between polyphenols (from grain and hops) and certain proteins from grain – a different class from the ones that are responsible for foam. And so, as we have seen in Chapter 5, the brewing process is geared to elimination of excessive levels of these materials. Excellent process efficiency is important also if other potential haze formers are not to survive into beer. These include starch, oxalic acid, and the polysaccharides from which the cell walls of barley are composed.

The color of beer is primarily due to so-called melanoidins, molecules that are not found in wine. These are formed in a reaction first studied by Louis Camille Maillard (1878–1936). The reaction involves a melding together of sugars and amino acids in the presence of heat. The greater the number of starter molecules and the more heating, the more color is produced.

For beer, the main stage for the Maillard reaction is malting. The sugars and amino acids made during the germination of grain are cooked together in the kilning stage. So, the more extensive the germination and the more intense the heating (to the extent of roasting in the production of specialty malts), the darker is the color. Not only that, the flavor is more extreme too, for flavorsome substances are produced in the Maillard reaction. Mild heating leads to the pleasing malty notes. Ale malts are heated more intensely than are lager malts, so the former tend to be darker and maltier. As the heating is prolonged and intensified, so we progressively pass through toffee and chocolate to the more acrid coffee-like characters. The dark malts concerned are the ones that are used as part of the grist for making porters and stouts. Just as for wines, color in beer can also originate in the polyphenols. This is much more significant for the very pale lager beers, as there is much less of the overwhelming melanoidin component. When polyphenols are oxidized, they darken – the self-same reaction is involved in the browning of sliced apples – and so any opportunity for oxygen pick up in the brewing process will increase the risk of color formation by this route.

Like foam, the color of beer has a profound effect on the perception of flavor. It is possible to make extracts of roasted malts that have been stripped of flavor but retain an intensity of color that allow them to be used to change the appearance of a beer from being lager-like to ale-like. Despite the fact that there is no impact on flavor as judged in tests where people can't see the product, when drinkers see the darker color they instinctively describe the taste and aroma as being like an ale and not a lager.

Some adjuncts are used to lighten the color of beer. The single biggest use for rice in the United States is for the production of Budweiser products. This is certainly not for cost reasons: The same care and attention goes into selecting the rice as is devoted to choosing the malting barley and, furthermore, rice needs to be cooked prior to use, which is an added energy cost. The rationale for using an adjunct such as rice is to lighten the color, but also to lighten the flavor, taking away some of the heavier nuances delivered by malt.

In terms of true flavor determinants in beer, then these are numerous. In one estimate, some 2,000 compounds are present in beer, whereas perhaps no more than 1,000 impact the character of wine. For beer, the flavor comes from the diversity of grist components: from hops, from yeast, from water, and, in the case of beers such as Lambic, from other microorganisms. And much more besides – fruits, herbs, spices, chocolate, the list goes on. For wine, we are talking just grapes, yeast, and a few other organisms, plus wood. And, well, there are beers aged in wood, too.

When we talk about "nosing," then the same technique, were it overtly applied in the world of beer, would be even more relevant. What one does is smell the substances that partition out of the liquid and into the "headspace" above. Various factors, such as temperature, impact this with more aroma being detected as the temperature increases. One very significant factor, though, is the alcohol content. Alcohol tends to reduce the extent to which molecules move into the headspace – so for most beers, there will tend to be a *greater* preponderance of aroma than for wine, because the majority of beers contain much less alcohol than does wine.

We have already encountered characters developed during the kilning and roasting of malt. To these we can add a molecule known as dimethyl sulfide (DMS), which plays an important role in the aroma of many foodstuffs. It is generally described by sensory specialists as being the classic aroma of cooked corn. However, it contributes heavily to the aroma of marine products, both seaweeds and animals such as crabs. It features in tomato ketchup. Most relevantly to us, it is a notable component of many lager beers, especially those produced by Germanic traditional techniques. Some brewers consider it a defect and strive to avoid its presence. Others like it at a low to medium level. Most hate it if it is excessive.

It is salutary to consider DMS in some detail, because it illustrates well the complexity of malting and brewing and how control needs to be exerted at diverse stages if the level of DMS is to be controlled to within desired limits. This example demonstrates the efforts that maltsters and brewers go to achieve *consistent* excellence in their products. It is not sufficient to allow nature simply to take its course and hide behind a shield of "vintage."

The precursors of DMS are developed in the malting process. The key one is called S-methyl methionine but we will call it SMM. It is made by the embryo during germination of barley, so the more germination occurs, the more SMM is produced. Furthermore, different varieties of barley have different capacities for making SMM. Other impacts on its level are the environment in which the barley is grown, how much nitrogen fertilizer is put on the field when the barley is grown, and how mature the barley is before it is malted.

Heating breaks down SMM, and so when the malt is kilned, the SMM is broken down to DMS and this is blown off by the stream of air on the kiln. Hence, the more strongly the malt is kilned, the more SMM is broken down and the less survives to enter into the brewery. Thus, malts destined for ales contain less SMM than do those intended for lagers, and as a result, ales generally contain less DMS than do lagers, because the SMM represents "DMS potential." On the kiln, some of the DMS produced is further converted into dimethyl sulfoxide (DMSO), of which more momentarily.

In the brew house, SMM is extracted into wort during mashing, but is broken down in the boiling stage. The more intensely the wort is boiled, the greater the breakdown of SMM to DMS which, in a vigorous boil, is driven off. So length and vigor of boil are very significant parameters. A brewer not wanting too much DMS will boil the wort more intensely – but he or she needs to be careful, because there is the danger of excessive color development by Maillard reactions and the development of cooked flavors.

The next stage is the whirlpool – and this is hot enough to continue SMM breakdown but there is not the same vigorousness and so the DMS released remains in the wort. So, a brewer wanting some DMS will temper the boil to allow significant levels of SMM to survive into the whirlpool stage to be converted to the desired level of DMS.

The next stage is fermentation and with it the huge production of carbon dioxide. This purges off much of the DMS – so the brewer wanting some DMS to survive will ensure that there is an excess quantity of DMS in the wort after whirlpooling. However, the DMSO kicks in here. It is converted by yeast into DMS to supplement the amount that emerged in the brew house. Different yeast strains produce different amounts of DMS. Fermentation at cooler temperatures leads to more DMS production. The amount of nitrogen available to the yeast for its growth modulates the amount of DMS produced. Finally, certain spoilage bacteria can produce copious quantities of DMS from DMSO.

So there we have it: one small molecule, one of thousands in beer, and one whose regulation depends on skill and attention to detail by farmer, maltster, and brewer alike. There is no equivalent of this precise and conscientious flavor control exercised in the world of wine. More so for beer than wine, there is the exertion of control over the consistency of the product as delivered to the consumer. Indeed, if it truly happened for wine, why smell the cork and sniff a soupçon in the glass? The very act of choosing the wine would have set you on your chance route, from which presumably there should be no turning back. You don't open up bottles of salad dressing or olive oil in the supermarket, just to check whether they are fit for purpose.

Water, as well as other process ingredients, provides salts to the beer, and these have a direct impact on taste. For instance, the chloride ion is said to provide mellowness to a beer and sulfate, dryness. Ergo, brewers control these levels. And they ensure continually that the water is free from any taints.

Hops provide bitterness and aroma. The level of bitterness is dependent on hop variety, both in terms of the absolute level of the resins that are the forerunners of the bitter acids in beer and with regard to the type of resins. There are three major resins and each generates two bitter acids. That's six acids in total – and they differ in their intensity of bitterness.

Conversion of the resins into bitter acids is called "isomerization" and this traditionally occurs in the boiling stage. Accordingly, the boiling conditions, both time taken and pH of the wort (which is determined by the grist and by the water), impact how much bitterness will be generated. And the hop variety used will also have a sizeable role.

The resins can be extracted from hops with liquid carbon dioxide and isomerized in a factory, separate from the brewing process, in which case they can be added to the finished beer to fine-control bitterness, or to generate some or all of the bitterness, or to convert a beer with a low bitterness to another with a higher one. Depending on how they are isomerized, there are different proportions of the six bitter acids produced, so this, in tandem with variety selected, means that different intensities of bitterness can be developed in products with ostensibly the same total bitter acid level. In other words, if there are two samples with identical concentrations of bitter acids, but one contains predominantly the more bitter types, then the *perceived* bitterness will be greater.

If you think that is complicated, then consider the oil component of the hops, which give the aroma to many beers. There are more than 300 different types of compounds in the oil fraction and a less-than-complete understanding of how the levels of each of them come together to deliver given hoppy noses. Much is done by expert selection of hops (rubbing and sniffing) and by the skill of the brewer in adding the hops at just the right time. Some varieties of hops are especially prized for their oil content in terms of amount but more especially the spectrum of aromas that they can deliver. All of the oils are volatile, so if the hops are added at the start of boiling, the oils are driven off and no hoppy character survives. So, for traditional lager brewing, a proportion of the hops is retained to near the end of boiling, enabling some of the oils to remain in the wort. Through the action of yeast, there is a modification of the compounds concerned, so that emerging in the beer is a complex yet generally delightful character known as "late hop." For traditional ale brewing, an even more robust technique is applied, so-called "dry hopping," in which a handful of whole hop cones is added to the finished product before it goes into a barrel.

And so to yeast. When it metabolizes wort, it produces a diversity of flavor compounds. In addition to DMS, it makes certain other sulfur-containing substances, including sulfur dioxide. It also makes higher alcohols and esters, which have fruity flavors, organic acids, fatty acids, and vicinal diketones. It is the latter that are generally considered reprehensible in beer and which need to be eliminated. Chief among them is a molecule known as diacetyl, which reeks of butterscotch and is added to popcorn to deliver the characteristic aroma of that foodstuff. Few of us want our beer to stink of popcorn (although it seems that it is a flavor accepted in – and, indeed, well suited to – Chardonnay, causing some California wines to be known as "butter bombs"). Diacetyl is naturally produced by yeast during fermentation, but the organism subsequently obliges the brewer by consuming it again. However, this takes time, and accordingly, the brewer is obliged to delay emptying the fermenter until the reprehensible aroma is dealt with. Another source of diacetyl is through the action of spoilage organisms, such as those that can inhabit beer dispense pipes.

There is really very little difference between the range of flavor compounds produced in wine and beer during fermentation. Nor would we expect there to be, for Saccharomyces is involved in each case. It is the fermentation conditions that have a profound effect on the levels of the various flavorsome substances produced. The factors include temperature, hydrostatic pressure, and the strength of the wort (in the case of beer) or must (in the case of wine). Generally, musts are stronger in terms of sugar content than are worts, and this tends to lead to high production levels of the fruity esters. But some beers are produced from exceedingly strong worts, so small wonder that these beers, which have an alcohol content similar to wine, also happen to have very wine-like flavors.

Most beers are relatively acidic, with a pH of around 4. The lower the pH the more sour is the product – and the most sour beers are those produced in Belgium, the so-called lambic and gueuze products. The high acidity is due to the action of a diversity of microorganisms that supplement the yeast in effecting fermentation. To temper this acidity, many of these beers have additions of fruit, such as cherries, black currants, raspberries, and peaches.

The wheat beers tend to have very high levels of fruity esters, but true Weissbiers also possess a clove-like character that arises through the metabolism of the strains of yeast that are used in the production of the most authentic beers of this genre.

Just like vintners, brewers talk of mouthfeel in beer. A major contributor is carbon dioxide, directly interacting with the trigeminal pain receptor system to afford "tingle." Conversely, nitrogen gas, which some brewers use to boost foam stability, offers a smoothness to the beers that contain it. Some firmly believe that the polyphenols afford astringency and texture, just as they do for red wine, but although the self-same molecules are present, they do tend to be in lesser quantities in beer than in red wine, if not white wines.

Before leaving beer quality, let us remind ourselves that, exactly as for wine, the glass into which it is introduced is very important for presenting the beer to its best aspect in terms of appearance and flavor. One of the most comprehensive studies of this factor was made in Belgium by Guy Derdelinckx. He employed fifteen trained tasters to assess some 400 beers in four different glasses: thistle, chalice, cylindrical, and cup-shaped. Among his findings were that Pilsners worked best in cylindrical glasses, while the high alcohol and fruity Trappist brews tasted best in wide-brimmed chalice or cup-shaped glasses.

8. Types of Wine

W INES CAN BE CLASSIFIED IN SEVERAL WAYS. THIS MAY BE according to their alcohol content, their color, or the amount of carbon dioxide that they contain. However, most frequently, they are grouped according to their geographic origin (Chablis, Bordeaux, Mosel, Chianti, and the like) or on the basis of the variety of grape from which they are produced. As we saw in Chapter 4, terroir and varietal and an interaction between the two have various degrees of impact on the result in the winery. A single variety grown in different specific locations within a single region may lead to differing end results.

In the United States, it is the norm to label wines on the basis of the grape varietal that enters into their production – Chardonnay and Merlot and Pinot Noir and Zinfandel, and so on. To be named in this way, the wine must feature more than 51 percent of that variety.

As we saw in Chapter 4, the prime species of grape employed worldwide is *V. vinifera*. Within this species are Muscat-flavored varieties, other varieties with their own distinctive flavors, and some with no distinctive flavors. In turn, within each of these classes are grape types that suit white or red wines. The Muscat varieties, with their characteristic aroma likened to daphne flower, are best suited to dessert wines.

Other *V. vinifera* grapes with distinctive flavors are, for reds, varietals such as Cabernet Sauvignon (with bell pepper and green olive notes), Merlot (green olive), Petite Sirah (tannic and fruity), Pinot Noir (peppermint), and Zinfandel (raspberry). For whites, we have the likes of Riesling (tart, fruity-floral), Chardonnay (apple), Gewürztraminer (spicy), Sauvignon Blanc (fruity, herbaceous), and Sémillon (figs). Those *V. vinifera* varietals without distinctive character are useful for bulk use.

Other species with some value are *V. labrusca*, with their foxy notes, and *V. rotundifolia*, which deliver very strong fruity characters. Finally, there are hybrid species, which hold some interest for the variations in flavor they may offer, characters variously listed as bitumen, bitter, horehound, and mustang.

As a primary categorization of wines, we can split them into the *table* wines, which are typically in the range 8–14 percent alcohol by volume (ABV) and the *dessert* and *appetizer* wines, which for the most part run at 15–21 percent ABV.

Considering first the table wines, they can be subdivided into the *still* and the *carbonated*. In the United States, a sparkling wine is defined as one that contains 0.392 grams (g) of carbon dioxide per 100 milliliters (mL) of wine. This is the equivalent of one atmosphere of gas. Champagne or "Champagne-type" wines typically contain in excess of 4 atmospheres of carbon dioxide, while "Pearl" wines contain less than 2 atmospheres. Sparkling wines tend to attract higher rates of taxation.

The most famous of all the sparkling wines, of course, is Champagne. Like the bulk of carbonated wines, it is white, another famous product being the Muscato Spumante wines from Asti, with the varietal being of course Muscat. There are pink Champagnes and sparkling Burgundies. Vinho Verde, from the Douro region of Portugal, undergoes malo lactic fermentation in the bottle or just before bottling. As a consequence, a significant level of carbon dioxide may remain in this wine.

The Champagne process is generally attributed to Dom Perignon (1668–1715), a cellarer in the Benedictine Abbey at Hautvilles, which is near Reims in France. It probably was a happenstance arising from the bottling of incompletely fermented wine. Reputedly the cleric was said to have exclaimed "Come quickly, I'm drinking stars!" Champagne is made from Chardonnay and Pinot Noir grapes. The base is known as *cuvée* wine and is usually relatively dry. It is mixed with sugar and yeast to yield the secondary conditioning matrix known as *tirage*.

In considering still wines, then, our next subdivision is into the dry, mellow, and sweet products. The first two of these are offered in white, rosé, and red styles, but sweet wines are invariable white or rosé. Some wines are flavored with additional components. Perhaps the best example is Retsina, a Greek dry white wine, produced usually from the Savatiano grape, with added pine resin.

Turning to the dessert and appetizer wines, again we can have whites, rosés, and reds. Amid the whites, we can include products such as Muscadet, Angelica, and herb-flavored wines, such as Vermouth. Pinks include tawny Port. Most Ports and many Sherries are red, but there are white versions of each. Dubonnet, too – a bittersweet fortified, wine-based aperitif that is flavored with quinine and herbs – is mostly available in red form but there is also a white version.

Angelica is a wine named for Los Angeles in California. Such wines tend to be fruity and very sweet owing to a large amount of residual sugar. They are typically made from Mission or Muscat grapes and the alcohol is enriched to 10–15 percent by using brandy.

White Port is similar – but possessed of less sugar, color, and flavor. White Ports are produced as for red Ports, though using white grapes.

Muscatel is a sweet wine made from Muscat-flavored grapes. Such wines are full-flavored and golden-brown in color. White dessert wines (except Muscatel, which may be aged for more than three years in oak) are consumed soon after packaging, as they are not intended for storage.

California Tokay is a pink dessert wine bearing no relation to the famous Hungarian wine of that name. It is produced by blending ruby Port, California Sherry, and Angelica.

Vermouth is a fortified white wine, of which there are sweet (red) and dry (white) versions based on Muscat, flavored with fifty or more herbs, barks (originally wormwood), and spices allowed to steep in the wine before separating prior to aging.

The chief fortified wines are Sherry, originating in Spain, Port from Portugal, and Madeira from the Portuguese archipelago of Madeira, which is 600 miles off the coast of North Africa. Fortification as a technique originated because the local soil and climate were not well suited to the production of grapes of inherent excellence. Enriching with alcohol also allows protection against microbial infection. Sherry and Madeira are fortified using spirit continuously distilled from the wine. Fortification of Port is with wine spirit.

Sherry is only made from white grapes, but Port and Madeira may be produced from either red or white grapes. Wines going to Sherry tend to be dry and fortification occurs post-fermentation. Any boosting of sweetness is through the addition of grape-derived products downstream, usually wines that have been fortified at the start of fermentation. If alcohol is added at the start of

fermentation, there is a suppression of yeast activity and, conse-
quently, sugars survive.

Port is usually fortified midway through the primary fermentation, and as a result tends to be sweeter than Sherry because unfermented sugars survive. Madeira may be fortified through either route
depending on the sweetness targeted in the product.

The wines used to make Sherry derive much of their character
from aging in oat "butts," but flor technology may also be involved
(see Chapter 4). In contrast, it is grape characteristics that are rather
more important for wines going into Port, especially the red variant.
Much of the character of Madeiras develops in the *estufagem* process,
which is a heating of the product.

Young un-aged Sherries are classified into either *finos* or *olorosos*
depending on their characteristics. Finos are dry, light, and pale
gold, not usually sweetened and have an alcohol content of 15.5–
17 percent. They are matured under flor yeast for three to eight years.
Olorosos, which are matured in the absence of flor yeast, are rich
dark mahogany wines with full noses, generally sweetened and with
alcoholic contents of up to 21 percent. The higher levels of polyphenolics in these wines suppress flor development.

Newly fermented Sherries are left to mature unblended for
approximately one year. They then pass to a blending process (the
"solera" system), in which the aim is to introduce product consistency. It comprises a progressive topping up of older butts of wine
with younger wines. A Sherry must be aged for a minimum of three
years before sale. During aging, flor prevents air from accessing the
Sherry, and so microbial spoilage and oxidative browning is pre-
vented. If there is no flor, as in olorosos, then oxidative browning can occur. Amontillado Sherries are produced with an initial
flor maturation followed by aging in the absence of flor, in which

oxidation reactions contribute significantly to character. They are also sweetened.

Red wines destined for ruby Port will have been aged for three to five years in wood. Those going to Tawny will have been aged in wood for more than thirty years. Vintage Ports are from wines of a single harvest that are judged to be of outstanding quality. They will be aged in wood for two to three years and then the aging completed in bottle for at least ten years. Ports are blended, especially the rubies.

Madeiras are mostly aged in wood. Vintage Madeiras must come from a single variety in a single year and must be aged for more than twenty years in wood and at least two years in bottle. Blending of Madeira is a simplified version of the Port system.

Marsala from Sicily was imported by English wine merchants at the end of the eighteenth century as a substitute for Sherry. Dry whites are mixed with grape juice and fortified to 18–19 percent before aging in oak casks and blending in a solera system.

Sangria is a deep red color and hence the name, which is derived from the Spanish word for blood. It is made by mixing red wine, fruit juices, soda water, fruit, and perhaps liqueurs, brandy, or cognac. Sangria blanco is made with white wine.

A Country-based Tour of Wines

Wines such as Sherry, Port, Sangria, and the like have clear national provenance, despite the fact that the styles have long been made in locales far beyond the countries of their birth. Thus, for instance, we have Californian and South African "sherries."

By far, the most common styles of wine, however, are the table wines, dare I say the "common or garden" reds, whites, and rosés.

As we have seen, they may be classified according to grape varietal, as is the case in, say, the United States. Equally, they may be classified according to country. Time, then, to fly through the world of wine, remembering that what follows is not intended to be comprehensive but rather indicative of what variation there is. At the outset, one might observe that diverse countries produce wine. Indeed, it is almost easier to name those that don't. My focus is on the main producers.

Ironically, London is generally considered to be the wine capital of the world, despite the fact that the United Kingdom is very much beer country and England is hardly prized for its wines. This may be unfounded: Champagne producers are actively courting vineyards in Sussex, which is less than ninety miles from Champagne, only 1°C cooler on average, and possessed of the same chalky soil as the classic Champagne region. There just happens to be the English Channel (or La Manche, if you're French) washing over the land in between.

Wine comes from all parts of France except the extreme north. The wines of France boast labels of quality, notably appellation controllées and vins delimités de qualité supérieuse, descriptors that speak to the pedigree of the wines and their production under strictly delineated conditions. Laws demand that traditional practices must be employed (analogous to the Bavarian brewing law, the Reinheitsgebot). For instance, the appropriate cultivars for a given region must be used. All of the grapes for Champagne must be pressed when they are still on the bunch and they must have been harvested manually. The list goes on.

There is fierce protection of names according to regions in France. *Alsace* was once part of Germany and this is reflected in the wine styles from this region. Just as in California, the labeling of wines in Alsace is according to varietals. This is a relatively cool locale and the most celebrated wines are some of the Rieslings and Gewürztraminers.

Burgundy is famed for the wines from its *Chablis* region, the grape of preference being Chardonnay. Because of less-than-hospitable growing conditions, the wines carry a pronounced acidity that is said to balance nicely the fishiness of marine cuisine. Other regions in Burgundy include *Côte d'Or*, where the grapes are grown on well-draining slopes. Here, it seems to be a case of Chardonnay for the best whites and Pinot Noir for superior reds. In the *Beaujolais* region, the preferred variety is Gamay, an early maturing grape that affords very fruity reds. The technique of Maceration carbonique that we encountered in Chapter 4 is employed in the production of Beaujolais Nouveau.

We move on to the *Rhone* and high-quality reds from the region near Avignon called Châteauneuf-du-Pape. And thence to *Bordeaux*, famed for its clarets beloved in England from the time of the marriage of the future Henry II to Eleanor of Aquitaine. Notable names include the red Médocs, the white Graves, and the sweet white Sauternes. Finally, we head to the *Loire* River and the Sauvignon Blanc of Sancerre and wines from Vouvray, some of which are slightly sparkling (pétillant).

German wines are usually lower in alcohol, at around 9 to 11 percent ABV. The grapes tend to be grown in cooler locations, which leads to a natural high acidity. Hence, acid reduction techniques such as malo lactic fermentation are applied, and sugar or grape juice is often added to musts. Notable wines are from the Rheingau west of Frankfurt and the region further west known as Moselle. The moniker for quality in German wines is Qualitätswein, with the appellation Qualitätswein mit Prädikat marking the best.

The best known wine from Portugal is Vinho Verde from the Minho region north of Oporto. The grapes are grown on trees and pergolas, which means that the grapes don't ripen extensively and are high in acid and low in sugar. This makes malo lactic fermentation essential and, as this continues in the bottle, there is significant carbonation. The alcohol content is less than 10 percent.

In Spain, the reds from Rioja are notable while in Italy the best known regions are Piedmont, Lombardy, and Tuscany. Tuscany is noted for its Chianti, especially the reds.

Australia has enjoyed a wine industry since the eighteenth century, in each region of South Australia, Western Australia, New South Wales, and Victoria. The wine industry in New Zealand is much smaller. Various countries in South America produce wine, perhaps most notably Chile.

And so to North America, where you will permit me to dwell in view of the provenance of this volume. The grape has long grown wild in the Mississippi valley, with the vines creeping up trees and yielding small, seed-filled berries with low sugar, high acidity, and characterized by a foxy flavor. The major native variety is the vigorously growing Concord.

The early settlers shipped better varieties to the New World, but these did not thrive in the harsh winters and humid summers, the latter leading to infection with mildew. The ravages of phylloxera ensued.

Vines first entered into California with the first missionaries from Baja California in 1769. The variety was Criolla, but this was later named Mission in honor of the trekking brothers.

The first Europeans arriving in California prior to the Gold Rush settled in the Los Angeles area. It was Jean Louis Vignes that delivered V. *vinifera* from France.

The Hungarian viticulturalist Agoston Haraszthy (1812–1869) moved up from San Diego to the Sonoma Valley to establish the Buena Vista vineyard in 1857 and he went on within twelve months to publish a pamphlet on enology and how to cultivate grapes for wine. Haraszthy persuaded California Governor J. G. Downey to sponsor a visit to Europe in which 200,000 cuttings of grape stock were collected.

Early California vineyards were cultivated by John Sutter, General Mariano G. Vallejo, the retired British serviceman J. H. Drummond, C. H. Wente, Charles Krug, and Gustave Niebaum, a Finnish sea captain who pioneered wine making in the Napa Valley through the Inglenook winery. It was soon realized that the best California wines were from the cooler coastal counties, Napa, Sonoma, and Santa Barbara. Nonetheless, some splendid wines come from the Central Valley.

It is salutary to take a look at some of the specifics of the California wine industry. Bottles retailing at less than $8 account for some 75 percent of the total volume but less than 40 percent of the total revenue. Those priced at over $15 per bottle amount to only 10 percent of the total volume, but a similar total revenue – and probably half of the total gross margin.

Serving Temperatures for Wine

Fuller bodied wines are properly served at somewhat warmer temperatures. Thus, a red wine warrants 65 to 68°F (18–20°C), a white wine 55°F (13°C), and a sparkling wine 50°F (10°C). Prior to serving, wines should be stored in cellars at 55–60°F (13–16°C), with the bottles laid on their sides to avoid air bubbles adjacent to the cork.

9. Types of Beer

THERE IS NO REASON WHY BREWERS COULDN'T PLAY ON THE strengths of grain or hop variety in the styling of their products. At present, it is fairly tangential – for instance, a brewer may boast about his or her use of hops with a certain provenance, perhaps Tettnang or Hersbrucker in Germanic lagers or Fuggles in English ales. Surely, this could be done much more: The characters unique to an aroma hop are no less special or unique than those associated with a given grape varietal. So where a winemaker may compare Pinot Noirs, Chardonnays, and Zinfandels produced by different vintners, so could the brewers, with no less sophistication or passion, debate the merits of different beers using Saaz or Challenger or Cascade.

The fact is they don't. Rather, brewers rejoice in styles that are founded on the type of grain, the extent to which the wort is fermentable, the strength, the color, and the brewing technique.

As a first fundamental, we speak of ales and lagers. The former have the more ancient pedigree. Once upon a time, the word "ale" was used to describe unhopped products. These days, virtually all ales are hopped. A classic feature of an ale is that it is fermented using a "top fermenting yeast," that is, a yeast that rises to the top of the vessel during fermentation. Fermentation is effected at a relatively warm temperature, say, 16°C (61°F), which leads to relatively high levels of esters (fruity flavors). Classically, the barley was well germinated during malting and then kilned to a relatively high temperature, conditions that caused a rich color and pronounced malty and toffee-like aromas. Finally, a small proportion of the hops are added to the fermented beer, giving a robust dry hop aroma.

The malts for lager-style beers are less well modified and more lightly kilned. Hence, they are lighter in both flavor and color. The yeast used is classically one of a bottom-fermenting type, in that it sinks to the base of the tank during fermentation, which tends to be at a relatively cool temperature (as low as 6°C/43°F) in which the yeast produces fewer esters but more sulfur-containing volatiles. The classic hopping method is to hold some of the hops back to later in the kettle boil, allowing some of the essential oils to be retained and these, perhaps modified by the action of yeast, cause a "late" hop aroma in beer.

Ales

When discussing ales, one predominantly has England in mind. Even though the last forty years have been characterized by a shift towards lager-style preferences in the British Isles, it is still ale that features as the traditional beer style of England and Wales. Scotland has embraced lager-style beers for far longer than has England, with brands such as Tennent's dating to 1885.

The classic English pale ale tends to have a copper color and to display a tremendous balance of malty and hoppy character. Neither is done to extreme, nor is the bitterness level. These beers tend to have clear fruity notes and alcohol levels of 4 to 5.5 percent by volume (wort of 11–14° Plato). When served on draught, classically delivered from casks using a hand pump (beer engine), a pale ale is known as a "bitter." Carbonation levels are relatively low (perhaps 1.2 volumes as opposed to the 2.2 volumes that might be present in a kegged version of the same beer, or 2.6 volumes in a bottle or can). Furthermore, the alcohol content in a draught version may be somewhat lower. Some brewers seek to deliver such low carbonation products in a canned form for home consumption. To promote foaming, in the absence of a hand pump they often resort to a widget in the can, together with a dose of nitrogen gas. This ruins the authenticity of the product, notably its hoppy character, especially as the beer must be kept at the excessively low refrigerator temperature if gushing is to be avoided.

Scottish ales tend to be sweeter and more toffee-like than English ones, and are traditionally divided into "light" and "heavy" depending on their alcoholic strength (2.8–3.5 percent vs. 3.5–4 percent). It will be noted that the term light here is entirely different from "light" beer (or the misspelled "lite"), which in an American context means low carbohydrate rather than low alcohol. Small wonder that light beers have failed in the United Kingdom whereas they are the biggest selling beers nowadays in the United States.

The classic English ale is of course the India Pale Ale (IPA). The genre is characterized by strong hop bitterness and higher alcohol content (5–7 percent) than "regular" pale ales. The first IPA was brewed by George Hodgson in 1822, close to St. Mary le Bow in east London. The Burton-based brewers had already established a fruitful trade with Baltic countries, shipping their beer by way of the Trent Navigation into the Humber and thence into the North Sea through Kingston-upon-Hull. But Napoleon's blockade and a severe Russian tariff on imports ate heavily into the business. It was Allsopp's of Burton, followed by Bass, who really took the IPA trade by storm. In those pre-Pasteur days, the very real concern was one of preservation. Hence, the very high gravities, alcohol contents, and the phenomenal bittering. It is often not realized that, while some bottled beer was shipped, by far, the bulk of the beer sold to India was in casks, for bottling locally. Hop bitter acids by no means kill all organisms, and the most prolific inhabitant of those casks bouncing on the ocean waves was Brettanomyces. The typical flavor notes produced by this organism are "barnyard" or "mouse pee." Modern-day IPAs, happily, lack this touch of authenticity, although there is at least one brewery in California that very much seeks to champion the action of this yeast.

Going out of fashion now in England are Mild Ales, which are very malty, low hop products of relatively low alcohol content (3.2–4 percent). Some of the milds are quite pale in appearance, but many are darker – a rich brown color, in which case the bottled versions are called Brown Ale. An Old Ale is reminiscent of a mild ale, more alcoholic (6–9 percent) and aged over yeast either in a tank in the brewery or in the bottle (bottle-conditioned).

Porters and stouts are ales. The advent of porter came in 1722 and owed its success to the burgeoning London populace. It is a beer born of the industrial age and it presaged the advent of the first mega brewers. It emerged as competition for the ales that were reaching the metropolis from towns such as Burton-upon-Trent. But it is a style shaped also by the laws of taxation in those distant days. Duty was levied on the basis of raw materials as opposed to the strength of the wort or beer. The darker brown malts kilned over untaxed wood were cheaper than the more heavily taxed paler malts that were cured on coal, which attracted hefty attention from the exchequer on account of the unhealthy impact of burning coal in urban environs. Thus, beer brewed from the paler malts fetched twice the price – tuppence. Richer people could afford ale – and rejoiced in its appearance as viewed through the newer drinking receptacles made from glass. Meanwhile, the London brewers met the budget of the industrial workers, including the porters in the city's markets, by producing large quantities of beer brewed from the coarser brown malts.

Some say that London water was too bicarbonated to make decent pale ales, but was satisfactory for darker products, which could tolerate the higher degrees of extraction from malt and hops that occurred under less acidic conditions. In the very dark, bitter, and richly flavored products, a multitude of sins could be hidden! Ergo, porter brewers were less rigorous in their demands for choicest malt and hops. Of course, plenty of the latter were added (hops were lightly taxed) making porter rather resistant to spoilage – and the relatively high alcohol content (we believe at around 7 percent ABV) would have further deterred marauding microbes.

This robustness meant that the beer could be brewed in bigger and bigger vessels, with the attendant economies of scale. Aging the beer in vast tanks (the legendary exploding Meux Horse Shoe Brewery vessel of 1814 supposedly held some 5,500 hectoliters) took the harsher edges off and afforded some clarification, but in truth clarity was not a major priority for the average industrial worker in eighteenth- and nineteenth-century London.

The advent of Combrune's saccharometer in 1762 meant that brewers could now see the much greater yields of extract that could be derived from paler malts. Hence, the switch from brown malt to a combination of pale and black malts that were now being produced in Daniel Wheeler's roasting cylinder. The net color of the beer changed: brown to black. And the "extra" dark (and stronger) variants were of course "extra stout porters," nowadays usually simply called stout.

Porters today span the spectrum from mid-brown to black. They may have significant but not excessive roasted notes, but retain substantial maltiness and sweetness and not excessive bitterness or hoppiness. Alcohol content is generally in the range 4.5–6.5 percent.

The great Irish-style stouts display much more roasted, harsh, coffee-like character. They are more bitter than the porters, but tend to have less hop aroma. They may have a slight lactic nature due to a deliberate tolerance of lactic acid bacteria. Their alcohol content can be as low as 3.8 percent but as high as 7.5 percent (Export-Style). Through the use of nitrogen gas, the foams are luscious and white, and the nitrogen has the impact of cutting through some of the harsher character that emerges from the burnt grist.

English stouts have a rather different character. The Imperial Stouts can range in color from dark brown to black and have an alcohol content anywhere between 7 and 12 percent ABV. They are very malty but not excessive in roast character. They are fruity and hoppy and very bitter.

Our son is now twenty-seven. His mother still recalls fondly the days after the birth when her batteries were recharged in the hospital with sweet stout, sometimes called milk stout, not necessarily because of its ability to promote lactation in the mother (though many believe that it does do that) but because of the whey-derived lactose employed in a production process dating back to 1669. Lactose is, of course, left alone by brewers yeast, and it has only one-fifth of the sweetness of sucrose. Apart from this modest contribution to sweetness, lactose also affords body. These beers are malty, but not roast, and of relatively low bitterness and alcohol content.

In oatmeal stouts, perhaps 5 percent of the grist comprises rolled oats together with some of the heavily roasted adjunct used in the production of the Irish stouts. They can run somewhat dryly on the palate.

The strongest beers from Britain are the barley wines, which may reach 12 percent ABV. Because the wort is extremely strong, the yeast tends to produce very high levels of the fruity esters and higher alcohols, which give vinous character. These tend to be the overriding feature of such tawny beers, although they do have some toffee maltiness and hoppiness.

The Irish have the red ales, which are analogous to the English pale ales, though more estery and less hoppy and, of course, possessed of a distinct reddish hue from the kilning process. In Germany, the ales pre-dated the lagers, and some significant ale styles remain.

Alt ("old") is a quite hoppy and bitter copper or brown crisp dry ale from the Düsseldorf region. Kolsch, from Cologne, is a paler, less hoppy relative. Both contain alcohol in the region of 4.5– 5 percent.

Most notable among the German ales are the wheat beers. Perhaps the best known are the weissbiers (weizenbiers) from Bavaria, served in tall tulip-shaped glasses and strictly *without* a slice of lemon despite what some believe in North America. Such distinctly turbid and straw-colored liquids are characterized by high carbonation, a distinct fruitiness, and a clove-like aroma and are traditionally served as a morning accompaniment to white sausage, sweet mustard, and Germany's gigantic equivalent of the pretzel. German regulations dictate that the grist must comprise at least 50 percent malted wheat, not raw wheat. Typically, the upper limit is around 70 percent because of lautering difficulties with a huskless cereal. It has been a Bavarian staple since the fifteenth century, when it was called weissbier (white beer) on account of its much paler appearance when compared to the dark ales that were the standard beer in that state 500–600 years ago.

This is arguably the healthiest of beers. The hefe weizens (hefe = yeast; weizen = wheat) are loaded with yeast (vitamins source). And they're as refreshing as any seltzer – such beers traditionally contain 3 to 5 volumes of carbon dioxide.

In fact, wheat beers presently account for roughly a quarter of all the beers brewed in Bavaria – something akin to the output of pilsner. Just as for other ales, the high temperatures and open vessels also encourage ester production.

In Germany, it is illegal to use a bottom fermenting strain for weizenbier production – so there are no "lager" wheat beers in the strictest sense of the term. *Hefe weizen* is the most popular style of wheat beer. Apart from the clove-like nose, it also will tend to be very estery, but low in bitterness and will be around 5 – 5.5 percent ABV. *Kristall weizen* is the bright equivalent, similar to hefe weizen except that it typically contains 1 percent by volume less alcohol and the color is kept low by employment of the palest of malts. Contrary to hefe weizen with its bottle conditioning, kristall weizen is bulk conditioned in pressure vessels prior to cold conditioning and filtration.

Dunkel weizenbier is similar to hefe weizen except that it incorporates highly kilned malts in the grist, though the color may also be enriched by the use of *farbebier*, coloring beer. *Weizenbock* has an alcohol content of 7–8 percent ABV while at the other pole is *Leichtes weissbier*, which is made from low-strength wort and develops an alcohol content of 3–3.5 percent ABV.

Napoleon's troops marauding in Northern Germany in the early nineteenth century called the local Berliner Weisse "the

Champagne of the North." This is an altogether different product from the Bavarian wheat beers. The alcohol content is even lower than for Leichtes weissbier at 2–3 percent ABV and the pH much lower (3.2–3.4) on account of a simultaneous lactic fermentation. The belief is that this style is at its peak when aged for at least eighteen months. It is deemed sublimely refreshing in the summer months – drunk from bowl-shaped glasses (as opposed to the tall weizenbier glasses) – with the tartness edged off with essence of woodruff, caraway schnapps, or raspberry syrup.

The Belgians have their own wheat beers (*witbeers* to the Flemish speakers, *biere blanche* for the French), which are very pale, use unmalted wheat, and are flavored with sweet and bitter orange peel and coriander. They are conditioned in the bottle and are very cloudy. Alcohol content is around 5 percent ABV and bitterness relatively low.

Nowhere else on earth has the diversity of beers been enjoyed as by the Belgians, each in its own glass. Flanders ale (Oud Bruin or Oud Red) is a richly colored brown ale that is rather sour, spicy, and fruity. Some of these ales have a woody nature. They all have modest bitterness and an alcohol content in the vicinity of 5 percent. Dubbel is a maltier, sweeter, and nuttier beer that may be up to 6 percent ABV while Tripel is spicy, estery, quite pale, and possessed of up to 8 percent ABV. Especially famous are the beers from the six Trappist breweries of Westmalle, Westvleteren, Chimay, Rochefort, Orval, and Achel, which between them brew some twenty very strong, bottle-conditioned ales.

Table beers, which contain between 0.5 and 3.5 percent ABV, are traditionally consumed alongside meals in Belgium by all members of the family. They are light in body and low in carbonation and bitterness, but relatively sweet and smooth.

Saison is a beer-style associated with smaller artisan breweries and it is sometimes called "country beer," as it is traditionally the drink of agricultural workers. As well as hops, other spices may be used in a beer whose versions can range from very pale to brown in color and from 4.5 to 9 percent ABV, indicative of the challenge of precisely defining such a product.

There is a range of other pale and strong ales, reaching up to 11 percent ABV, but no beers are more unique to Belgium than the Lambic and Gueuze products, which traditionally are made in the French-speaking Brussels region of the country – if made elsewhere they must be called Lambic-style or Gueuze-style. Lambic is a French word, so make that Lambiek in the Flemish. Here, again, is a beer with wheat, perhaps 40 percent of the grist being that cereal in its unmalted form. This is not the source of the challenge. That comes in the complexity of the fermentation, with its plethora of microbial contributors. Saccharomyces of various types is aided and abetted by the likes of Pediococcus, Lactobacillus, Brettanomyces, Candida, Hansenula, Pichia, and goodness knows what other bedfellows. The result is a beer of genuine complexity, and flavors akin to "wet horse blanket" and "mouse pee" as well as profound sourness. They are even more complex when rendered as a blend of young and old ferments and accompanied by additional fermentation in bottle (gueuze). Ergo, and analogously to Berliner weisse, we find the introduction of fruit – cherries for kriek, raspberry for framboise, peaches for peche, and black currants for cassis.

Lagers

Up to the sixteenth century, all German beer was ale. Furthermore, lager as most now know and believe they recognize it, that is, a pale or golden and bright refreshing liquid, dates back only some 160 years or so.

The origin of lager (the word meaning "to store") can be traced to 1553 when summer brewing was outlawed in Bavaria on account of the increased risk of infection in the warmer months. The brewing season was between St. Michael's Day (September 29th) and St. George's Day (April 23rd), ergo the storage of beer in cold caves to survive the summer months. Hence, too, the advent of the *Marzen* beer style: darkish brown, high in alcohol, and rather hoppy to afford preservative value. It was mashed-in during March, stored through the summer, and consumed as required. Any left over before brewing commenced again in September was disposed of in that most joyous of parties, the Oktoberfest.

Consumers even in those days evolved increasingly sophisticated tastes, and when the glass emerged as the preferred type of vessel from which to imbibe, brighter and more golden liquids were perceived as being more appealing and refreshing. Progressively Marzen, and the other dark lager style *Dunkel*, lightened in color. The development was primarily on account of Gabriel Sedlmayr at the Spaten brewery in 1871, who had been impressed by colleague Anton Dreher's recipe developed in Vienna of 1841. But it was the British who overwhelmed the pair of them, and one of the techniques they had learned on their 1833 study tour of England and Scotland was how to make paler malts.

The German word for pale is *Helles*, and thus we have the workhorse style within the lager genre. It is traditionally brewed with relatively hard water, and I have heard it said that this is the reason that hopping should not be overdone or otherwise the bitterness becomes overbearingly harsh. This whole business of the ideal water for different beer styles is firmly placed in the dogma file: would that there was a body of unequivocal scientific evidence to justify some of the claims made for the impact of parameters, such as hardness, carbonate, sulfate, and chloride. Whatever the truth is, helles traditionally is moderately hopped, of course, with varietals that provide that desirable late hoppiness, such as Tettnang and Hallertau.

Helles is one style in every self-respecting lager brewer's portfolio. Another, of course, is *Pilsner* (German *Pilsener*), a style originating in the Czech (erstwhile Bohemian) city of Plzeň. That ancient place is close to the Zatec region (German Saaz) and hence the classic hop of that name is employed to impart both the bitterness and the late hop character that should be found in this beer. The water of Plzeň is soft, as it happens, and first found its way into the great golden nectar in 1842 under the stewardship of Bavarian Josef Groff with the aid of yeast spirited from the fatherland. What is the difference between Pilsner and Bock? The purist would say it's the malt to hop balance, with the Helles having the more pronounced malt and the Pilsner the more hoppiness. The philistine (pragmatist?) would say it's to do with whatever you make the label say. There is nothing that mandates the materials or brewing techniques that one is obliged to employ in the production of these or any other lager style (leaving aside general considerations such as the Reinheitsgebot).

Marzen, Helles, and Pilsner are classically in the 4–5 percent ABV range. *Bock* is stronger, perhaps 6–6.5 percent ABV, a May-brewed example being, well, *Maibock. Doppelbock* is stronger, perhaps 6.5–7.5 percent ABV or more, though not as strong as *Eisbock*, surely the original ice beer (it supposedly dates from the late nineteenth century) with an alcohol content of 9–11 percent ABV. Eisbock is dark, whereas Doppelbock can come in pale or dark versions, the latter of course traditionally delivered by the use of specialty malts. Perhaps here we can identify the most substantive difference between dark lagers and dark ales or stouts: their degree of harsh, burnt, acrid, coffee, and chocolate character. Whereas a stout grist would include sizeable amounts of roasted barley and roasted malts, the high color in a Doppelbock or a *Schwarzbier* (4.5–5.5 percent ABV) will come from the substantial use of Munich malts and the very sparing use of the heavily roasted adjuncts. And if it's genuine smokiness you want, then seek out *Rauchbier* with its pronounced character from the "bacony" or peated Scotch stable.

In North America, the main beer styles are subtly flavored lagers. The reader should not linger under the impression that these are somehow inferior or easier to produce. The more gentle the flavor of a beer, the more readily do off-flavors reveal themselves, for example, during storage. These beers are especially suited to consumption alongside hotter foods and in very warm climatic conditions.

Although in decline in the United States, variants of these styles have been ice beers and dry beers. In the production of the former, there is some degree of ice formation and removal during processing, leading (it is claimed) to a smooth crisp flavor and a slightly higher alcohol content. Dry, as a style, remains popular in Japan: The term is exactly analogous to its use in the world of wine and indicates a very comprehensive fermentation. This leads to beers with crisp, clean finishes and little aftertaste.

Light Beers

Four of the top five domestic brands in the United States are light lagers. It all started in Brooklyn in the sixties with the Rheingold breweries' Gablinger's. Brew master Joe Owades recognized that some people were fighting shy of beer either because they didn't like the taste (solved these days by "malternatives," of which more later) or because they feared that it would make them fat. So he added an extra enzyme to his brew, amyloglucosidase, which allowed all the starch to be converted to ethanol during fermentation, and the process was completed by restoring the increased alcohol content to a "normal" level by adding water. The result was a beer within the typical alcohol range, but possessing less residual starchy material and therefore lower calories. It also had relatively less flavor and foam potential because that final dilution means fewer malt kernel equivalents per glass.

The technology is still used to this day for some beers of this genre (others employ strategies involving replacement of malt with more fermentable adjuncts and ensuring that brew house conditions are such as to maximize the opportunity for the malt enzymes to act). The problem with Gablinger's was its marketing, which positioned it as a healthy alternative to the customer's usual beer. It was the equivalent of saying don't eat that nice juicy steak, take a lettuce leaf instead, it's the healthy thing to do: not what people wanted to hear.

However, the ball was on the roll, and next to come to the plate was the Peter Hand Brewing Company in Chicago, which developed Meister Brau Lite. It seems that they were as unsuccessful as brewers as they were incompetent at spelling, and they duly folded, the brand name passing to Miller, newly acquired by Phillip Morris. Now the marketing was subtle and successful: Here was a beer just as good as other beers; it just happened to be less filling. They brought in a top football running back to wave the can around and Miller Lite was rolled out on a national platform in 1975. By 1990, it had more than 10 percent market share. And it had been joined by Coors

Light, Natural Light, and, in the 1990s, Bud Light, which is now the number one beer brand in the United States.

Low-Alcohol Beers

For many beer aficionados, the term low- or non-alcoholic beers represents a non-sequitur. When judged as a non-alcoholic beverage, many of them hardly pass muster flavor-wise when compared with the obvious competition (colas, orange juice, cold tea, water, etc., etc., etc.). With the exception of beers such as the lower alcohol table beers in Belgium and the Berlin weisse products, most of the beers in the low alcohol genre seem to be little more than ersatz products for those pretending to enjoy a full-strength drink in the eyes of their peers. Personally, I no more see the need for a low-alcohol beer than I do the need for decaffeinated coffee or meat substitutes: If you don't want alcohol, then don't take a beer. If you're avoiding beef, then have a nut cutlet. If you're cutting out caffeine, drink water.

Only one thing would change my mind on this, and that is if somebody developed an alcohol-free or non-alcoholic product of genuine quality. I have yet to be convinced that anybody has. The simple reason for this is that alcohol makes a direct contribution to quality. For a start, you can taste it – it gives a warming and, well, alcoholic note. Just as importantly, though, it also impacts on the aroma delivery from other components of the beer. The balance of esters, other alcohols, sulfur-containing substances, and so on, with respect to the extent to which they leave the beer and hit the naso-olfactory system, is very much impacted by ethanol.

This is the reason why alcohol-free beers made by stripping out alcohol and then blending back a cocktail of flavorsome materials do not taste the same as their full-blooded beer counterparts. Take out the alcohol, rip out the heart of a beer.

There are those who fall back on more spurious arguments for not taking such a product. Take the Church of the Latter Day Saints, for instance: "Even if it doesn't have alcohol in it, drinking beer (and the atmosphere you would most likely be in) will keep the Spirit from being with you."

(As a good Episcopalian, I would venture to suggest that the Spirit would not be with you for the simple reason that there isn't a decent amount of alcohol in the drink, but let's not get hung up on matters sacred.)

Little Alternative

There has to be a genuine societal reason for going with an alcohol-free beer. Something like Prohibition, perhaps. The rationale doesn't apply in a country like Saudi Arabia, where beer has never been on the menu. But imagine those victims among our forebears who had to endure Prohibition in a beer-drinking country such as the United States. One alcohol-deprived product at the time was Bevo from Anheuser-Busch. Forced with the decision of whether to throw in the towel or make the best of a pretty bad job, they developed products of this type to tide them over until common sense prevailed once more. Presumably, Bevo was little more than unfermented wort, far-sightedly invented three years before national Prohibition came in 1919. It became comfortably the most popular of the diverse "near beers" of the age and, at its zenith, it sold some 5 million cases annually. It boasted that it was "The All-Year-Round Soft Drink. Appetizing – Healthful – Nutritious – Refreshing. Milk or water may contain bacteria. BEVO never does." Quite how many people added sugar and yeast to it and waited for something more interesting to develop is not documented.

Four years before the end of Prohibition, however, sales had flattened to 100,000 cases, and production ceased. All that remains by way of a reminder is a district of St. Louis called Bevo.

The 1980s was the only era when there was a seemingly genuine desire to explore the low-alcohol beer genre. According to U.K. legislation, to be called alcohol-free, a product had to contain less than 0.05 percent alcohol by volume, or in other words, an ethanol content not dissimilar to orange juice. The moniker non-alcoholic demanded less than 0.1 percent ABV. Low-alcohol essentially covered those products below 1.2 percent ABV.

One thing that we quickly realized technically was that relatively little alcohol goes rather a long way. Thus, a 2 percent ABV beer was vastly, vastly better than was one at 1 percent. However, in the United Kingdom, which is where I was working on developing such products, taxation issues held sway. Below 1.2 percent meant zero excise duty, but every increment above that made for a progressively more expensive product. And so the vector was unavoidably towards inferior products. Put another way, Her Majesty's Government were fiscally dissuading brewers from producing compromise strength brands that had lower than average levels of alcohol but were genuinely potable as beers in their own right. Looking back from the early years of the twenty-first century, when the British government speaks of binge drinking being a British disease, there seems to be a certain irony here.

For reasons already stated, there really is no entirely satisfactory way to make a low-alcohol or alcohol-free beer. The first option is to not ferment at all, or at best just "wave" the yeast at the wort under very cold conditions for a short period. Such products invariably have the authentic "brewery taste," by which I mean they taste like a brewery when wort is on the boil. Do you really want your beer to smell that way? It's a bit like suggesting that your steak-alternative should taste like a farmyard or an abbatoir.

High-temperature mashing may restrict the extent to which fermentation can occur. Mash in at, say, 72°C and the starch is largely converted to non-fermentable dextrins, so you can leave the yeast as long as you want on the wort (to clean up some of the rawer edges). This is about as good as it gets for a low-alcohol product – which isn't saying terribly much. And you can strip the alcohol from the product by reverse osmosis or low temperature vacuum evaporation, adjusting the product post-event with your proprietary blend of aromas.

High-Alcohol Beers

If one is derisory about the alcohol-challenged products, then does it follow that beers replete in alcohol are at the other pole of acceptability, that is, the ultimate in beer excellence? What are we to make, I wonder, of a beer released in recent years in the United States that boasts 25 percent ABV and retails at around a hundred dollars a bottle? It is the latest in a long line of macho products from one company, a succession that has taken in beers of 11 percent, 17.5 percent, 21 percent, and 24 percent ABV.

Such "big products" are pushed as post-prandials, which is no surprise, considering they are port- and liqueur-like in their potency and their aromatic character. Two ounces is the recommended dose, from one of only 8,000 bottles brewed, the latter being the shape of a brew kettle. To produce these very high-strength products is technically demanding and one surmises the "tricks" include the batch-dosing of more and more sugar throughout the fermentation and the use of very alcohol-tolerant yeasts.

Stepping back one rung on the ladder of alcohol madness we find the so-called "super lagers" of the United Kingdom. Such products are of a comparable strength to wines at around 10 percent ABV. A couple of cans of super lager are tantamount, then, to a bottle of wine. It might be argued that they should be considered on the same playing field, sipped in limiting and shared volumes alongside food. If such were a legitimate stance, then one might fairly expect to see the beers raised to a higher sophistication quotient, in premium packaging. An opportunity, if they are not saddled with such a gung-ho descriptor as "super." As it is, such beers tend to be pitched at a "lowest common denominator" level.

In the United States, in the 6.2–7.5 percent ABV range, we find the malt liquors. The term is suffused with confusion: To some, any beer is a malt liquor if it is brewed from malt. Usually, though, it is used to describe lager-style products that are more alcoholic than mainstream beers, and in some states, the presence in a product of more than a certain amount of alcohol may lead to the requirement that the product is not labeled as a beer but rather as a malt liquor. These products are often sold in wide-necked (for easier drinking) forty-ounce bottles as opposed to the customary twelve-ounce (350-mL) bottle or can. There is little escaping the image that these products have as the drink of choice for the down at heel.

What Else?

There is no limit to the ingenuity of brewers. In the United Kingdom, I have had oyster stouts, which were formerly beers to enjoy alongside oysters, but nowadays might be products that contain oyster essence. Another well-loved stout is chocolate stout, so named for its content of chocolate as well as chocolate malt. The British have long since enjoyed Shandy (ale blended with lemonade), and other citrus-containing beers are available worldwide.

There are many beers that are brewed with spices such as coriander; compare, for example, the gruit of the Middle Ages. The Mexicans love their Michelada, a mix of lager with lime, salt, Worcestershire sauce, soy sauce, and Tabasco sauce. A recent launch in the United States was a beer incorporating ginseng, caffeine, and guarana. Pumpkin beer has long been popular.

There are beers brewed with honey as part of the grist. There are gluten-free beers, some of them made from sorghum. There are beers aged in wood, like a wine or a whiskey. Then, of course, we must mention the mixed drinks. Stout plus bitter ale yields Black and Tan while Black Velvet is stout and champagne.

A last category is the low-malt "beers" produced in Japan. Several years ago, Japanese brewers spotted a loophole in the country's taxation laws, which indicated that if a product contained less than 25 percent malt, then it would attract far less taxation than the norm. As a result, there was a burgeoning of products in which the malted barley was heavily substituted by adjuncts. The products could not be labeled "beer" and they were clearly inferior to the conventional products, but this was more than compensated in the eyes of the consumer by their much lower cost. Now the Japanese brewers have gone a step further, realizing that products with no malt in them at all attract even less taxation. Thus, we have the "third" category of beers, based on a diversity of curious ingredients that include soy- and pea-derived material.

Serving Temperatures for Beer

The gently flavored American-style lagers are best consumed at essentially refrigerator temperatures (0–4°C, 32–39°F). Products with fuller flavors warrant perhaps 8–12°C, (45–54°F), with 12–14°C (54–57°F) for many ales and as high as 16°C (61°F) for some of the heavier, stronger ales.

Malternatives

Finally, let us dwell briefly on malternatives, sometimes called "alcopops" or flavored alcoholic beverages (FABs). These are basically drinks of a similar alcohol content to mainstream U.S. lagers (around 5 percent ABV) but which do not taste of beer, but rather of whatever flavor is added to them, say, orange, lemon, cherry, or cola, and so forth. They can (depending on the state) be taxed as beers if they are made from a beer (unhopped) that has been decolorized by passing through activated carbon. These products have attracted a lot of adverse publicity because they are seemingly positioned at a young (but legal) market. They tend to be sweet, and the human only develops a taste for more bitter products with age. Died-in-the-wool brewers decry them because they perceive their popularity as meaning that it will be increasingly less likely that drinkers will gravitate to "real" beers.

10. The Healthfulness of Wine and Beer

O N NOVEMBER 5, 1991, A JOURNALIST BY THE NAME OF
Morley Safer presented a piece on the highly rated CBS news
program, *Sixty Minutes*, that told of how a French doctor, Serge
Renaud, had shown that red wine was a powerful counter to coronary heart disease. The phenomenon became known as the French
paradox, for the French were pursuing a diet that was anything but
healthful and seemingly certain to cause serious blocking of the
arteries, but in fact that wasn't occurring. Almost overnight, sales
of red wine started to surge in America and the wine industry in the
country took massive advantage.

It was far from being the first report of alcohol consumption being
beneficial to health, as we found earlier in this book when discussing
the history of wine and beer. Indeed, the Irish doctor Samuel Black
in 1819 remarked on the far lesser incidence of angina in France as
opposed to his own country on account of "the French habits and
modes of living." The first properly scientific study in this area was
conducted by the Baltimore biologist Raymond Pearl in 1926, all the
more remarkable for its timing in that it was smack in the middle of
Prohibition. I suppose it was based on data accumulated before the
Volstead Act, which enabled federal enforcement of Prohibition.

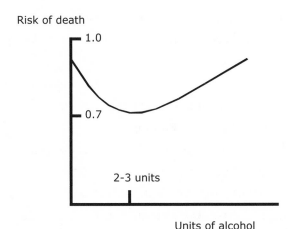

Figure 10-1. The impact of alcohol consumption on the risk of death. One unit equates to a glass of wine or a 12-ounce serving of regular beer.

Pearl reported that those who drank moderate quantities of alcohol lived longer than those who were total abstainers, and thus was demonstrated the U- or J-shaped curve that has now come into common usage (Fig. 10-1).

Since Pearl's time, there have been numerous reports to substantiate the claim that drinking in moderation is beneficial, notably insofar as it substantially reduces the risk of atherosclerosis ("hardening of the arteries"). Considering that more than 15 million deaths annually may be attributed to this cause, the effect seems eminently worthy of investigation. To cite just one such piece of work, in data from an American Cancer Society study, it was claimed that the risk of death dropped by 16 percent for those claiming to take one drink per day, and even at six drinks per day, the risk of death from this cause was lowered by 8 percent.

In atherosclerosis, there is a restriction of blood flow due to the accumulation of fatty materials (atheroma) on the walls of the arteries, leading to strokes, heart attacks, and death. It would appear that the consumption of red wine counters this through lowering levels of so-called bad cholesterol and lessening the tendency of blood platelets to aggregate.

For long enough – and still to a significant extent – the claim was that the beneficial impact was unique to wine, especially reds. Thus, Dr. Selwyn St. Leger, working in 1978 out of Cardiff, Wales, concluded that for men aged 55 to 64, there was a clear benefit to drinking wine, whereas the poor men inhabiting countries where beer was the drink of choice had the highest rates of death from coronary heart disease. The clear inference was that there must be some component of wine, especially red wine, that is the "magic" factor.

David Kritchevsky and Davis Klurfeld, researching in Philadelphia, showed in 1980 that rabbits fed with a North American diet developed alarming symptoms of atherosclerosis, and that red wine, but not beer, countered this effect. Soon the theory emerged, largely on the basis of early work from scientist colleagues in my own university, that red wine was especially rich in antioxidants, notably certain polyphenols, including resveratrol, that have a profound effect on ensuring that bad cholesterol does not exert its negative impacts.

In actual fact, one needs to be very cautious when relating studies made in the test-tube to effects on the living body. It is important that the antioxidants are demonstrated to actually enter the body and reach the parts that matter. For instance, it has been suggested that although the antioxidant levels in red wine are at substantially higher levels than they are in beer, the large size of many of them makes for poor absorbability. In contrast, it was proven by Ghisselli, Natella, and Guidi that the blood plasma does actually contain the antioxidants after somebody has consumed beer, ergo they *are* getting into the circulation. Denise Baxter, working with colleagues from Guys Hospital, London, demonstrated that one antioxidant, ferulic acid, penetrates the digestive system far better from beer than it does from tomatoes. I daresay if I went down to my local Farmers Market with a bottle of beer in one hand and a tomato in the other and asked which was healthier, then most folks would opt for the red orb. In fact, at least for this antioxidant, they'd be wrong.

In 1996, a group of eminent doctors made a detailed analysis of the literature up to that date on the relative effect of wine, beer, and spirits in reducing coronary heart disease. These medics were Dr. Eric Rimm and Dr. Meir Stampfer (Harvard School of Public Health), Dr. Arthur Klatsky (Kaiser Permanente Medical Center, Oakland), and Dr. Diederick Grobbee (Erasmus University Medical School, Rotterdam). Their finding? "We conclude that if any type of drink does provide extra cardiovascular benefit apart from its alcohol content, the benefit is likely to be modest at best or possibly restricted to certain sub-populations." So now the American Heart Association says, "There is no clear evidence that wine is more beneficial than other forms of alcoholic drink."

John Barefoot of Duke University and Dr. Morten Gronbaek of the Institute of Preventive Medicine in Copenhagen emphasized in 2002 that the benefits of wine were more likely to be a function of the healthier lifestyle of those preferring that beverage as opposed to those who prefer beer. Wine drinkers consumed less saturated fat and cholesterol; they smoked less, and exercised more. It was the abstainers that had the worst habits, eating less fruit and vegetables, but more red meat, and they also smoked more. When connections between socioeconomic status and beverage preference were controlled for in the studies, it was shown that wine drinkers with the same income and social standing as beer consumers or abstainers had healthier lifestyles.

In 2006, Gronbaek analyzed 3.5 million checkout receipts at supermarkets in Denmark over a six-month period. Those buying wine bought more fruit and vegetables, poultry, low-fat cheese, and milk. Those buying beer also bought more sausages, chips, sugar, butter, and soft drinks. In other words, it seems that beer drinkers need to do more to improve their life choices. It seems that beer is no less beneficial than wine – with tongue-in-cheek we might wonder how really unhealthy beer drinkers would be if they did not take advantage of the beneficial effects of beer in reducing atheroma.

In fact, the evidence for the benefits of beer (in moderation) has been building. It is more than twenty years since Richman and Warren in a Canadian health survey reported that beer drinkers have significantly lower rates of sickness than the general population.

Turning to Klatsky and his colleagues again, they reported in a 1997 paper published in the *American Journal of Cardiology* a study of 3,931 people that showed that the weakest inverse correlation between alcohol consumption and coronary heart disease was for spirits drinkers. For men, the relationship was significant for beer, for women it was significant for wine. And it didn't matter whether that wine was red or white. So Klatsky wrote in 2001: "[I]t seems likely that ethyl alcohol is the major factor with respect to coronary heart disease risk. There seems to be no compelling health-related data that preclude personal preference as the best guide to choice of beverage."

One wonders, too, whether the impact might even be an indirect one. T. J. Cleophas of the Albert Schweitzer Hospital in Dordrecht, Holland, concluded that there is a significant psychological component in the beneficial relationship between moderate alcohol consumption and mortality. Is it alcohol itself having a calming influence, or even the deceleration in the pace of life that is associated with taking a drink, that achieves benefit by de-stressing the drinker?

Dr. Cynthia Baum-Baicker, a clinical psychologist within the University of Pennsylvania Health system, has reviewed the literature on the positive psychological benefits of moderate alcohol consumption and concludes that there is reduced stress in such consumers. There is an increase in happiness, euphoria, conviviality, and pleasant and carefree feelings but a decrease in tension, depression, and self-consciousness. More so, low doses of alcohol improve certain types of cognitive performance, such as problem-solving and short-term memory. By contrast, heavy drinkers and abstainers had higher rates of clinical depression than did regular moderate drinkers. Guallar-Castillon and colleagues at a Madrid university described a study showing that people drinking wine or beer *believed* themselves to be healthier; in fact, the higher the consumption the better people felt they were!

Perhaps this is the reason why studies in different regions variously flag up wine or beer as the more beneficial. If there is a major psychological component, and the well- being is linked to the euphoria induced by one's favorite drink, then this may explain why beer came out superior to wine in studies on the countering of coronary heart disease in beer-drinking societies, such as Honolulu, the Czech Republic, and Germany. For example, research from Czechia suggested that the lowest risk of heart attack was in men who drink between 4 and 9 liters of beer per week. Indeed, Dr. Hoffmeister of the Freie Universitat Berlin suggested that if European beer drinkers stopped taking their favorite beverage, there would be a decrease in life expectancy of two years, and a lot of unhappiness. By contrast, in studies in the wine-drinking rural environs of Italy, it is wine that is flagged up for its benefits.

And so it seems to be the alcohol that is the key ingredient. Still, though, there continue to be those who point to the very high levels of antioxidant in red wine, concluding that it must therefore unequivocally be better for the body. Those believers might fairly turn to the work of John Trevithick from the University of Western Ontario, who has shown that despite the higher levels of antioxidants in red wine, when the equivalent amount of alcohol in the form of beer or wine is consumed, just the same level of antioxidant gets into the blood from each.

In passing, we might also note the seeming importance of frequency. Dr. Mukamal and his colleagues at the Division of General Medicine and Primary Care of the Beth Israel Deaconess Medical Center in Boston found evidence that the most favorable impact of an alcoholic beverage, whether wine or beer, occurred if it was taken moderately and frequently. Better to have one or two daily then one or two on just a couple of days per week. And certainly better than storing up your week's intake in order to binge at the weekend.

Circulatory problems are not the only ailments seemingly alleviated by the consumption of alcohol. Andrea Howard from the Montefiore Medical Center is one among several to draw attention to the U-shaped relationship between the consumption of alcohol and the risk of diabetes, with an approximately 30 percent reduced risk of diabetes arising from moderate alcohol consumption. Considering that diabetes cost the United States some $132 billion in 2002 and is a leading cause of kidney failure, amputations, blindness, and cardiovascular disease, then the merits of moderate alcohol consumption again come to the fore. And, here again, the evidence is that it does not matter one jot whether the alcohol is taken in the form of beer or wine.

Guenther Bode and colleagues from the Department of Epidemiology at the University of Ulm found that moderate beer consumption countered the organism *Helicobacter pylori*, a risk factor for stomach cancer and the causative agent of stomach ulcers. Tero Hirvonen and colleagues from the Department of Nutrition of the National Public Health Institute in Helsinki found that a bottle of beer daily reduced the risk of kidney stones by 40 percent. Stampfer and his colleagues showed that both wine and beer reduced risk of gallstone disease. It seems that alcohol speeds up the emptying of the gall bladder after a meal. Frequency of intake was important: better to have five to seven days of moderate consumption rather than one or two. Holbrook and Barret-Connor from the Department of Community and Family Medicine at our San Diego campus showed that social drinking is associated with higher bone mineral density in men and women, in other words, a reduced risk of osteoporosis. At another California university, namely the Keck School of Medicine at the University of Southern California, Paganini-Hill cited alcohol as one of the aspects of leisure activity that served to counter Parkinson's disease. Researchers at the Bispebjerg Hospital within the University of Copenhagen showed that consumption of alcohol led to reduced instance of thyroid enlargement.

Veronika Faist and her colleagues from Garching, Kiel, and Munster found that some of the materials formed in specialty malts when they are roasted boost the activity of enzymes in the body that deactivate unwanted toxins. Tagashira and his team at Asahi Breweries found that the polyphenols from hops inhibit streptococci and delay the development of dental caries. Meanwhile, Nakajima's lab at Ochanomizu University in Tokyo showed that there was material found in darker beers that blocked streptococci from making a polysaccharide that they use to stick themselves to teeth.

One concern of many is that the excessive consumption of alcohol will, in the words of Shakespeare in *Macbeth*, "provoke the desire but take away the performance." There is no evidence that wine or beer is more or less impacting in the male response when consumed to excess. Furthermore, the indications are that alcohol must be abused substantially if there is to be anything other than temporary interference with sexual potency and performance. As E. M. Jellinek said, "Germ tissue could be damaged by very high concentrations of alcohol, but it is so wonderfully protected that before such concentrations of alcohol would occur, the alcoholic father or mother would be dead."

Moving on nine months, then, there is a mixed press about the merits of beer for the breast-feeding mother. On the one hand, Koletzko and Lehner from the University of Munich suggest that a barley polysaccharide promotes prolactin secretion, this being the pituitary hormone that promotes breast activity, while on the other, Julie Mennella from the Monell Chemical Senses Center in Philadelphia suggests that alcohol consumption might actually somewhat reduce milk production. All I can offer is experiential evidence: My wife Diane gave birth to our son in enlightened days in England in 1980, and very much enjoyed her Mackeson stout in the maternity ward after Peter emerged. It relaxed her.

In fact, there is substantial evidence that the drinking of alcohol helps mental processes. Christian and colleagues from Indiana University studied 4,739 sets of elderly twins to demonstrate a J-shaped curve between alcohol consumption and cognitive function. Those drinking moderately fared better than abstainers and those partaking to excess. Japanese researchers in the Aichi Prefecture

indicated that moderate drinkers had higher IQs than did abstainers. The type of alcohol in these studies was irrelevant, as indeed,
it was Ruitenberg from the Erasmus Medical Centre in Rotterdam,
who found that one to three drinks per day reduced the risk of
dementia. Cupples found that one drink per day for women or two
for men reduced the risk of Alzheimer's disease, while Dufour and
colleagues from the National Institute on Alcohol Abuse and Alcoholism in Rockville, Maryland, demonstrated that beer stimulated
appetite and promoted bowel function in the elderly.

As Professor Robert Kastenbaum of the School of Human Com-
munication at Arizona State University is quoted as saying: "There
is by now sufficient information available to indicate that moderate use of alcoholic beverages is pleasurable and beneficial for older
adults."

One undisputed fact is that drinking alcohol promotes the desire
to pee. Perhaps the most curious study was that of Nagao and friends
at Kyoto University, who found that the freshest beers are the ones
with the greatest tendency to give you the urge.

Another fear that is widespread is that alcohol consumption will
substantially increase the risk of cancers. In fact, the evidence suggests that alcohol consumption needs to be really rather substantial
for it to be a causative factor. And thus, the 1989 report of the Com-
mittee on Diet and Health of the National Academy of Sciences
cautions against *excessive* consumption of alcohol. There is in fact
a diverse and contradictory literature in this entire area, and even
some papers that argue that drinking wine and beer can *counter* certain cancers. An international collection of pharmacologists, epi-
demiologists, and toxicologists authored a letter in the *British Journal of Cancer* in 1993 in which they reported a critical examination
of the extant data on cancers of the mouth and gastrointestinal tract

and concluded that two drinks per day for men had actually halved the risk of cancer as compared to non-drinkers. It took a dozen or more drinks per day to raise the risk of cancer over that of abstainers.

Nevertheless, within the World Health Organization, the International Agency for Research on Cancer has classified alcohol as a group one carcinogen. Yet the National Institute on Alcohol Abuse and Alcoholism (NIAAA) draws attention to the tremendous inconsistency in the data reported. They caution against abuse of alcohol, and also draw attention to the claims that alcohol potentiates the damaging impact of the prime carcinogen, notably tobacco.

The National Cancer Institute in the United States says that alcohol consumption above the recommended daily intake increases the chances of developing cancer of the mouth, esophagus, pharynx, or larynx. But, then again, the NIAAA draws attention to the fact that although a few studies link chronic alcohol consumption to stomach cancer, the majority of investigations find no association.

Exactly the same conclusion has been made by the NIAAA for cancer of the pancreas. Indeed, one study has suggested that there is actually a reduced risk of contracting this devastating carcinoma if you drink moderately.

It appears that liver cancer is induced by the cirrhosis caused by abusing alcohol, rather than the drink per se. Unlike most of the other cancers of the gastro-intestinal system, in the case of cancer of the colon and rectum, there is a greater conviction of a link with alcohol consumption. In the words of the NIAAA: "Epidemiologic studies have found a small but consistent dose-dependent association between alcohol consumption and colorectal cancer, even when controlling for fiber and other dietary factors. Despite the large number of studies, however, causality cannot be determined from the available data." In other words, again, we see that the causative agent may not be drinking as such, but rather other elements of the alcohol-consumer's lifestyle.

When it comes to breast cancer, one of the most highly publicized of all carcinomas, there is very much a mixed bag of evidence. The NIAAA has concluded that *chronic* alcohol consumption increases a woman's chances of developing the disease by about 10 percent. Others find no increased risk, certainly not from moderate drinking. The influential Framingham Study drew the conclusion that moderate alcohol consumption does not increase the risk of developing breast cancer.

Indeed, here we must consider matters holistically, as surely should be the case in all investigations into these sensitive topics. For it has been claimed by the Mayo Clinic that folate counters the risk of contracting breast cancer – and we should all know by now that beer can be a plentiful source of this vitamin.

The Fred Hutchinson Cancer Research Centre says that the consumption of four or more glasses of red wine each week reduced a man's risk of developing prostate cancer by 50 percent, whereas beer and white wine were without impact, either positive or negative. Recent scientific reports are more favorable for beer, though, with Anna Harris and her colleagues in New Zealand and Australia reporting that PSA (an antigen that is diagnostic of prostate problems) was proportionately decreased by an increase in a man's beer intake.

The consensus of expert opinion, based on data to hand, is that there is no reason to associate the moderate consumption of alcoholic beverages with the risk of leukemia, cancer of the thyroid, lung cancer, bladder cancer, and skin cancer. In regard to cancer of the uterus and non-Hodgkin Lymphoma, the indications are that moderate alcohol consumption *reduces* the risk. Dispassionate consideration of all the above, then, surely leads us to the conclusion that beer and wine are not seething pools that will wreak havoc as they course through your body. In fact, they might even have meritorious components.

And so work, especially from Japan, in recent years has drawn attention to the presence of some interesting molecules in beer that have the potential to protect against the ravages of cancer. Arimoto-Kobayashi, of the faculty of pharmaceutical science at Okayama University, finds molecules like pseudouridine in beer capable of counteracting the influence of mutagens. In fact, he finds that pseudouridine only accounts for 3 percent of the antimutagenic impact of beer – so a rich seam appears to be un-mined at present. Perhaps we need to be looking at some of the interesting polyphenolics molecules, including those such as xanthohumol, isoxanthohumol, and 8-prenylnaringenin from hops, championed by scientists at Oregon State University. However, let us inject a note of caution. You really do need to consume rather a lot of beer to get truly helpful levels of some of these components – and what you reap on the swings of anti-carcinogenic benefit you will lose on the side of obesity and beyond.

Cancer is a touchy topic and not to be trivialized. Brewers and winemakers should be defensive to an appropriate degree: It truly does seem that drinking of beer and wine within recommended limits (as, for example, advocated by the United States Department of Agriculture [USDA]) is *not* going to substantially increase your risk of developing cancer. In fact, it might even improve your chances of avoiding some of the cancers. However, despite the song and dance surrounding the hop factors that may be potent anti-carcinogens, we really must wait for the dispassionate accumulation of evidence.

To overtly market a beverage product as "anti-cancer" constitutes stupidity.

Empty Calories?

Go to <http://www.student-manual.com/living/nutrition.htm> to find a lie. For there we find the "Unofficial Student Manual" in which prospective freshmen are given the advice "Empty Calories – Beer, candy, pizza, soft drinks are empty calories. It simply means your body gets nothing but calories and fat. In other words, you get hungrier despite gaining weight...stay away from them." I think I know the true purpose of these sentiments, namely, to warn young folk off alcohol. My personal preference is to find ways to illustrate to students (and those of legal drinking age) the very real pitfalls of abusing alcohol. But I simply will not tell untruths, and the notion that beer of all beverages constitutes "empty calories" is abject nonsense, as a glance at Table 10-1 will show. The data is taken from the USDA web site (<http://www.nal.usda.gov/fnic/foodcomp/cgi-bin/list_nut_edit.pl>). The reader is cautioned that there will be significant ranges to be found between different brands. The site tells that the beer approximates data drawn from the analysis of ales, lagers, porters, premium beers, and stouts, but the rest hinges on lager samples alone.

One element not listed in the table is silicon, yet beer is now accepted as one of the most significant sources of silicon in the diet. Jonathan Powell and his Cambridge University colleagues have amply demonstrated the benefits of moderate beer consumption in countering osteoporosis, in part, because of this high silicon charge.

Despite the clear evidence that beer (and to a lesser extent wine) do not comprise empty calories, there remains a tendency for some

Table 10-1. The Relative Nutrient Composition of Beer and Wine

Nutrient	Units	White table wine (one 5-fluid-ounce serving)	Red table wine (one 5-fluid-ounce serving)	Beer, regular (one 12-fluid-ounce serving)	Reference daily intake (males 19–50)
Proximates					
Water	g	127.7	127.1	327.4	3.7 liters per day
Energy	Kcal	122	125	153	1800
Protein	g	0.1	0.1	1.64	56
Fat	g	0	0	0	
Ash	g	0.29	0.41	0.57	
Carbohydrate	g	3.82	3.84	12.64	130
Fiber	g	0	0	0*	38
Sugars	g	1.41	0.91	0	
Minerals					
Calcium	mg	13	12	14	1000
Iron	mg	0.4	0.68	0.07	6
Magnesium	mg	15	18	21	350
Phosphorus	mg	26	34	50	580
Potassium	mg	104	187	96	4700
Sodium	mg	7	6	14	1500
Zinc	mg	0.18	0.21	0.04	9.4
Copper	mg	0.006	0.016	0.018	0.7
Manganese	mg	0.172	0.194	0.028	2.3
Selenium	μg	0.1	0.3	2.1	45
Vitamins					
Vitamin C	mg	0	0	0	90
Thiamine	mg	0.007	0.007	0.018	1.2
Riboflavin	mg	0.022	0.046	0.086	1.3
Niacin	mg	0.159	0.329	1.826	16
Pantothenic acid	mg	0.066	0.044	0.146	5
Vitamin B_6	mg	0.074	0.084	0.164	1.3
Folate	μg	1	1	21	400
Vitamin B_{12}	μg	0	0	0.07	2.4
Other					
Ethanol	g	15.1	15.6	13.9	

The dietary guidelines of the United States Department of Agriculture (<http://www.health.gov/DIETARYGUIDELINES/dga2005/report/HTML/D8_Ethanol.htm>) recommend a maximum of one drink per day for women and two for men, a drink being a 12-ounce serving of a regular beer or 5 ounces of wine (12% alcohol by volume). And it is stated that at this level, there is no association of alcohol consumption with deficiencies of either macronutrients or micronutrients and, furthermore, there is no apparent association between consuming one or two alcoholic beverages daily and obesity.
*Recent data shows that there is fiber in beer.

people in their ignorance to decry beer in particular. Even worse, some dietary fads have unfairly pilloried beer. In the South Beach Diet, Agatson declares that there are "good carbs" and "bad carbs," the former being less digestible carbohydrate polymers that stop insulin resistance and cure cravings. Bad carbs have the contrary effect, it is claimed. Agatson suggested that maltose is the worst carbohydrate of all and made the ludicrous statement that, because beer contains maltose, it should be avoided, this particular sugar being responsible for the "beer belly." The premise is naïve in extremis for the simple reason that the vast majority, if not all, of the maltose is consumed by yeast during fermentation. Although the inaccuracies have been remedied on the diet's web site, the claims damaged sales of "standard" beers, at the same time repositioning certain light beers, leading to the emergence of "low-carb" beers (must deliver less than 7 grams of carbohydrate per serving) and boosting advertising opportunities for spirits.

In reality, many wines contain more sugar than does beer, but the main source of calories in beer and wine is the alcohol itself. There are many conflicting opinions concerning what this means for weight and whether there is some mystical reason why calories in the form of alcohol should be more fattening than those in the shape of protein, fat, or carbohydrate.

But what about the beer belly? Wannamethee and Shaper of the Department of Public Health at the Royal Free Hospital School of Medicine in London conclude:

While metabolic studies indicate fairly unequivocally that alcohol consumption even in moderate amounts contributes to weight gain, the epidemiological evidence on the relationship between alcohol intake and body weight based on cross-sectional studies is conflicting.... Findings...suggest that light to moderate drinking is not associated with weight gain but that heavier levels (>30 g alcohol per day) contribute to weight gain and obesity in men and women. Overall evidence from prospective studies supports the concept that alcohol is a risk factor for obesity, as one might expect if the energy derived from alcohol consumption was added to the usual dietary intake.

One factor that must be carefully considered for those counting calories is the volume they consume. Some people are overwhelmed by the sheer volume of beer and prefer their alcohol in shorter, more concentrated doses. Thus, perhaps they may have two glasses of wine but just one serving of beer. For a wine of 100 calories per serving, this means that they are ingesting 200 calories, whereas they would be consuming just 150 calories from a regular beer, and only 100 from a diet beer.

Resveratrol, still believed by some to be an active ingredient in making red wine the best bet for countering atherosclerosis, reared its head again in the issue of countering obesity. A study at the Harvard Medical School and the National Institute on Aging found that mice dosed with resveratrol were better able to overcome the negative impacts of a high-calorie diet. The slight problem is that for the same benefits to apply to the human when taking into consideration the amount of resveratrol found in red wine, the wine aficionado would need to consume somewhere between 750 and 1,500 bottles of plonk a day.

Incidentally, Sonia Collin and her colleagues from the Catholic University of Louvain have found resveratrol in hops. No longer can the wine folks claim it as their molecule alone! One last point of difference between beer and wine: The latter invariably contains added sulfites, whereas beers, particularly in the United States, do not.

11. Conclusions about Beer and Wine — and the Future

J OE WAS ONE OF MY BEST EVER STUDENTS. HE THIRSTED FOR knowledge of beer. He devoured the literature, attended class compulsively, asked perceptive and passionate questions, and was superlative in the experimental brewery, producing beers of sublime excellence. He also has the word "beer" tattooed across an ample belly in huge gothic script. This says everything really. I cannot imagine the winemakers on campus – seemingly as many women as men – adorning their guts with "Pinot Noir," even in henna. Perhaps a subtle little tattoo located somewhere tasteful and sophisticated.

Plenty of these winemakers take my beer classes. They accept the well-intentioned teasing, for the most part (as do the chemical engineers, who proudly identify themselves when I go round the class at the start of term, before I remind them that they have no soul and have no grasp of the beauty inherent in biological systems). After about four weeks, as we finally reach yeast and fermentation, I delight in telling the viticulture and enology students that had this been a wine class, we'd have reached the fermenter on day one: "Heck, you just crush a few grapes and you're ready to toss in the yeast – if you bother even to do that. You might always just leave the contaminating microflora to get on with it." It is, of course, in good-willed jest, but with a straighter face I do remind the students that there really is a far more complex series of stages that a brewer (and before him or her, the maltster) must conduct with precision and consistency before the noble Saccharomyces comes to the party. And this, for the biggest brewers, occurs 24 hours per day, 365 days a year. Now *that's* a crush.

Yet, wine and winemakers have stolen the moral high ground. Turning up at a prestigious dinner on campus, I go to the bar. "What do you have?" "Red or white wine, sir." "What beers do you have?" "None, sir – but we do have sparkling water."

A year or two back, I showed up at a dinner at the home of one of the university's major officers. I asked why there was no beer, only wine. The reply was "but this is a culinary evening." The reality is that beer is probably just as worthy a match for foods as wine – but more of that momentarily.

What I strive for is greater reverence for beer in countries like the States. It exists in pockets but certainly nowhere near to the extent that it does in some other countries. Take Belgium, as we saw earlier. The right beer in the right glass of the right shape with the right emblem. *That* is giving beer its rightful treatment.

Yet, so often in the world, the treatment of beer is criminal. Venture into a restaurant in this neck of the woods and you will be lucky to get a glass if you order a beer, the bottle being unceremoniously plunked down on the table before you. If the glass does come, then the server will assiduously endeavor to pour the product with painful restraint, decanting the liquid laboriously down the side of the glass in an attempt to avoid generating head. I lose track of the times that I tell them that the foam is integral, nay critical, and that they should direct the stream vigorously to the bottom of the glass so that I can admire the glorious whiteness of the head. But then, as often as not, if a glass does appear, it has clearly been washed badly. There will be ugly great bubbles on the side of glass, evidence for fatty deposits.

But when it comes to wine, there is a wholly different scenario. "A bottle of the chateau *Exhorbitante de Profite*, sir? Certainly, an excellent choice." Moments later, along comes the wine waiter with said bottle, white towel over his arm, and, with great deliberation, he presents the label for your affirmation that you really have got the remarkable good taste to have selected this majestic beverage. The foil will be scored and removed and the cork pulled with due ceremony, and placed lovingly before me. With a flourish, a morsel of the nectar will be tipped into the glass and the sommelier will stand back, expectantly. I hold the glass to my nostrils and savor the bouquet (of course, knowing the ritual, I never taste) before announcing that all is well and my friends and I can take receipt of this gift from the gods. The bottle, if the wine is white, will be plunged into a gorgeously created ice bucket and the waiter will depart with an envious "enjoy!"

Each and every bit of that could be done with beer, but no. You are just as likely expected to chug it from the bottle. One is never invited to gargle in the same way with Mouton Rothschild Chateau.

It is all theater. For wine, it is Drury Lane. For beer, it is small town rep. And the reality is that beer is, to my mind and that of any genuinely discriminating and knowledgeable bon vivant, just as much Olivier or Dench as is wine.

Beer and Food

The widely accepted truism holds that it is red wine with dark meat, white wine with poultry and fish – and beer with pizza. Thankfully, there have been a series of studies recently that have debunked the whole notion of ideal pairings of alcohol with certain foods. In particular, there has been scrutiny of wine–cheese couplings. It has been suggested that the appropriate wine for a given cheese is one from the same country or region. Perchance there is sense in this, not only for wine and cheese, but for any food–drink pairing. Perhaps processes of natural selection at the dinner table narrow down the food and drink offerings to those most suited one to another. And thus, for beer, we might have Pizza and Peroni, Thai red curry and Singha, roast beef and Bass.

In the case of wine and cheese, another truism reveals itself in that the intensity of flavors in the two should be in balance. Not a cheese of rank rancidity with a delicate little white. Not something bland with a tannic, woody red. Here, again, we can surely extrapolate to any food–drink partnership.

Marjorie King and Margaret Cliff looked at pairing nine Canadian cheeses and eighteen wines to find that whites paired with a wider range of cheeses than did reds, and that the stronger the flavor of the cheese, the more difficult it was to find a match. Most telling, King and Cliff reported that individual judges varied greatly in their individual assessments, reflecting the significance of personal expectations and preferences. It was also emphasized that their study did not embrace other food components: Few of us subsist on a diet of cheese and wine alone, least of all brewers. We ought not to lose sight of this – a given beer or wine may not necessarily meld with all the items in some chef's culinary masterpiece. Finally, King and Cliff say, "Individuals should be encouraged to experiment to determine their own preferences."

One of my friends and colleagues, Hildegarde Heymann, used a trained panel of tasters to assess the flavor of eight wines before and after tasting various cheeses. She found that attributes such as astringency and oak were substantially decreased following the consumption of cheese. Only one aroma was enhanced – naturally enough, it was buttery. In other words, the net effect of the cheese was that it *diminished* the flavor impact of the wine, when the aim should surely be food and drink that *complement* one another. Hildegarde concluded that one could probably enjoy any preferred cheese with any preferred wine.

That there is a physiological as much as psychological component at play is clear. Our bodies do tell us what they need. If we are running too high on salt, either within a food or because we have been losing water through exercise, then the signal is to dilute with water. There are cells in the hypothalamus at the base of the brain that monitor the sodium and potassium levels in bodily fluids. If they run too high, then the thirst sensation is triggered. The same applies with the dryness of foods – it's a case of water, please. And surely that means beer more often than it does wine!

All of us have several thousand taste buds, each of them a mass of cells and nerve fibers. We understand that there are proteins on the surface of the cells that interact with different flavor-active molecules, leading to a change in shape of the protein and a change in the permeability of the membranes that in turn generate an electric current sensed by the nerve fibers. Off goes the message to the brain, which duly tells us whether we are sensing bitter or sweet and so on.

The majority of the flavors in food of course are not detected overtly as taste, but rather as aroma, through the nose. The olfactory cells are themselves nerve cells, ending in cilia that basically wiggle around in viscous mucus (yum) to detect aroma molecules. Once again, the theory is that molecules fit into specific receptor sites in a membrane.

Of course, every food contains a mixture of molecules that will react with the taste buds and olfactory cells. And so some substances will tend to interfere with the binding of other compounds, whereas others seem actually to promote the sensation derived from other elements of the diet. Thus, cynarin from artichokes blocks the sour, salt, and bitterness receptors, making everything taste sweet. Extracts of the Indian plant, *Gymnema sylvestre*, block sweetness. Alternatively, the legendary umami substance, monosodium glutamate, boosts the intensity of saltiness and bitterness. Components of smoked fish seem to boost the metallic character derived from beer.

We must also take into consideration the time for which a food and beverage are in contact with taste and aroma receptors during the dining experience. Might a drink wash away molecules from taste receptors? Conversely, there can be accumulations of materials that block the perception of other flavors, such as perhaps fatty coatings, which interfere with the binding of polyphenolics to proteins that registers with you and me as astringency or bitterness, depending on the size of the polyphenol. Equally, if there are polypeptides in a food that will bind polyphenols more rapidly than they bind to the detector portions in the mouth, then they would be expected to tone down the astringency of a high polyphenol beverage. Therefore, the proteins in cheese moderate the astringency of red wine, cider, or a tannic beer. There might even be reactions taking place between food and drink, leading to new flavor-active substances.

Alas, so much in this area is simply not worked out physiologically, psychophysically, or organoleptically. We resort to experiential, even apocryphal, learnings.

Some seem logical to me. Thus, a light-flavored drink complements a gentle-tasting food. If we are talking beer, then an example might be smoked salmon and a light American lager. Conversely, steak and kidney pie with a draught English ale, perchance. There may even be a visual phenomenon at play – the rich browns of a crusty meat pie demand at the very least rich redness or brownness in a beer. By the same token, an ashen white fish dish seems better suited to a delicately hued lager.

One might expect, too, some complementarity between a food and a beer for the key flavor notes within them both. Few of us like overt contrasts: A really bitter beer simply does not work with an overpoweringly sweet dessert. But some contrasts and complements do work, and consideration of some of our ingrained biases might help here. Thus, many folk have applesauce with pork, so pork and fruity beers as a pairing? And just think of Maltesers, known as malted milk balls in the States. Does this speak to us of chocolate and real malty, lower bitterness, lightly hopped ales?

Debbie Parker at Brewing Research International (BRI) looked at the impact of dry roasted peanuts on the flavor profile of an ale. The nuts seemed to tone down the malty and hoppy character, while rendering the beer sweeter and less bitter. Equally, the nuts became less dry and salty – but I suppose you might have expected this, on account of the liquid intake.

Parker suggested that the acidity of beers "cuts through" oiliness, butteriness, and creaminess. Meanwhile, she indicates that saltiness in foods will accentuate polyphenol astringency. In a study of the co-consumption of chicken fajitas and an unspecified style of beer, the BRI finding was that expectedly there was a decrease in oiliness and saltiness of the food, with a boosting of the onion, chicken, and pepper aromas. There was also, of course, a cooling of the spiciness. Not mentioned in this study, but perhaps of significance, is that the degree of carbonation might well be important here. Carbon dioxide reacts with the self-same trigeminal nerves that interact with chili – perchance they compete, meaning that a highly carbonated beer might tone back the fire more than a low carbonation product. Pure conjecture on my part – but so often that is the case in food–drink pairings.

Let's not try to shoehorn food and beer pairings. Some that I encounter as I travel the world are just silly. It really is a case of what works for you. Stout and cornflakes for breakfast may be your bag, but I don't personally recommend it. Let's just remember that there is absolutely nothing peculiar about wine that makes it a preferred accompaniment to food. In fact, the rehydrating properties of beer probably make it even more suitable for many dishes. But then again, let's be careful, for the same applies to water.

Beer for the Right Occasion

It is simply fact that there are more drinking occasions for beer than there are for wine. You may or may not agree with me that beer is just as frequently the right match for food as is wine. But few surely could suggest that wine would outstrip beer after exercise. Or in the sports stadium. Or when one is fishing.

And it is simply fact that there are vastly greater choices for types of beer than there are for types of wine. For non-fortified wines, we are basically speaking of red, rosé, and white, with alcohol contents from somewhat less than 10 percent to 15 percent or slightly more. Sparkling or not. And flavor derived from grape, yeast, bacterial action, and wood.

For beers, though, we have every color, from water clear to black, and alcohols from essentially zero to 24 percent plus, and flavor emerging from malt, other cereals, yeast, other microflora, water, hops, wood, fruit, chocolate, spices, vegetables, guarana, caffeine, ginseng...the list goes on. We have diverse carbonation levels; we have nitrogen gas. And what is more, brewers have control. Few and far between are the beers that vary significantly from batch to batch. There is no concept of vintage.

Which is not to say there couldn't be. It would be entirely plausible for brewers to relax their guard and allow for intentional change in the final package (remembering that there are some beers that are already marketed deliberately in this way, notably naturally conditioned products). My thesis is that this could be done with all beer, making it a part of the mystique. And thus, exactly analogously to what happens with wine, a beer would be allowed to change and then perhaps be prized for the flavors that it develops. "Hey, come over, I've got a 1991 Schlitz that you simply must try. It has a divine sherry note, suffused with black currant bud and moist straw, with just a soupçon of damp parchment."

Nonsense? Of course it is – but it is nothing less than what happened in the world of wine, a world in which individual bottlings are tolerated for their differences year on year. It is called vintage. Woe betide the unfortunate brewer if a senior manager in a quality conscious company detects that a batch of beer is not the perfect match for the brand concerned. Excuses such as, "Well, it's not been a good year for hops," do not flow.

In Chapter 1, we came across the man who bemoaned the fact that brewers strive for consistency and predictability. Whether it's a book having all its pages in the right order, or an automobile's brakes working every time you press them, or an airplane leaving on time, you expect quality and performance. So why do you tolerate variation in a wine? How has it happened that the whole concept of vintage and premium, nay ludicrous prices, has arisen?

In the Eye of the Beholder

There is no doubt whatsoever that wine is perceived as an altogether more sophisticated beverage than is beer. Alas, the brewers have shot themselves in their big feet on this one. Too often, their advertising has involved lavatory-level humor. It doesn't help anyone for microbreweries to imply that the big brewers are little more than factories (it is simply not true – the big companies set a gold standard on quality that some of the small guys would do well to emulate). And when it comes to one major brewing company attacking the products of a more successful competitor, then who does one believe? The answer is that Joe Public doesn't trust anyone in the beer business (at least on a large scale) and turns to wine, where nobody lambastes another and everybody laughs his or her way to the bank.

One of my students, Christine Wright, has delved into the mindset of consumers. She asked folks who were visiting a winery in Napa and those on beer tours in a California brewery to rank drinks for their healthfulness. There was an overwhelming belief in the superior health benefits of red wine, and beer was perceived as being nowhere near as good for you. But then Christine pointed out a few facts about beer, observing that it was at least the equal of wine in terms of its benefits in countering atherosclerosis and that it actually contained more nutrients than did wine, such as vitamins and fiber. The rating for beer improved – but was still ranked after wine, whether red or white.

It seems that popular yet ludicrously inaccurate messages hold sway, sentiments such as:

Avoid white wine, spirits, and worst of all, beer. (*The South Beach Diet*)

As for alcohol, everyone knows it makes you fat, especially beer and hard liquor. (*Suzanne Somers' Fast and Easy*)

I think of beer and wine as liquid sugar. (*Your Last Diet! A Sugar Addict's Weight-Loss Plan*)

In fact, it seems that the healthfulness, purported or otherwise, of a beverage is not the primary consideration when one is buying drink (and in the latter part of the study, Christine reached right across the United States). In decreasing order of significance were taste, what the activity being undertaken was, location, time of day, price, feeling afterwards, food pairing, company, serving size, alcohol content, weather, healthfulness, calories, and carbohydrates. Those questioned came out with statements like:

- The healthiness of beverages has little impact on my beverage choices.
- I drink the alcoholic beverages I like and do not worry about their nutritional values.
- I am loyal to my favorite brand or type of alcoholic beverage.
- I generally try to follow a healthy and balanced diet.

Yet, when questioned about the composition of beers, subjects were frighteningly ignorant. Few knew that there are vitamins and antioxidants in beer. And far too many believed that some things that are decidedly *not* found in beer in significant quantities are there, such as fats, trans fats, artificial sweeteners, preservatives, and artificial colorings.

There is a big education job to be done on beer. And it isn't the brewers that can do it, or dare I say, university professors perceived as having a personal axe to grind. The simple fact is that wine is perceived as being somewhat superior.

It can't be because its creation is more technically demanding – it isn't. The route from barley, water, and hops to beer is vastly more complicated than is that from grape to wine. Indeed, I well recall being asked to write an article on what the brewer can learn from the winemaker and asked help from a buddy who is a professor of enology for, as I said to him, "I can't think of anything."

As we have seen, wine is not better for you. There are no sustainable arguments to say that wine goes any better with food than does beer – in fact, I would suggest, for example, that virtually any Asian food pairs vastly better with a lager-style beer than it does with wine.

And it is not because wine has a higher intensity or complexity of flavor. I can pour for you a high-alcohol beer and you will detect exactly the same nuances of aroma that you will in a wine – with aromas from malt and hops thrown in for good measure. Don't forget that there are some 2,000 compounds in beer and only half as many in wine. In truth, it is like trying to compare tea and coffee. They are clearly different, but not to that great an extent. Thus, they are both decidedly wet.

We are talking about imagery. Napa and Sonoma are vastly prettier and altogether sexier than Milwaukee and St. Louis. Wine is presented as the *New Yorker*, but too often, beer is *The National Examiner*. We are talking *New York Times* meets the *New York Post*. It's *Masterpiece Theater* versus *Men Behaving Badly*. It is pretension versus pragmatism.

All in all, the wine folks have done an amazing job, and brewing and too many associated with it have screwed up. Beer is more technically demanding, more scientifically replete, healthier, and more complex. Yet, it is *perceived* as being inferior to wine. Happily, there have been moves to redress the balance, witness the Here's To Beer campaign. But there is still much to do.

Clues

There are clues out there for how brewers may do things differently. One clear example is the value of heritage and national provenance.

Some while ago, a colleague looked at people's preference of domestically brewed beers and those that were imported. When subjects were shown the bottles of beer that they were consuming, with the labels proudly displayed, then there was a sizable groundswell in favor of the imported products, the view being that they were somehow better. However, when beers were tasted "blind" and the drinkers had no idea which beer was which, then there was no significant preference. The beers brewed in the United States were deemed the equal of those imported. I haven't done the study, but I would lay odds on the fact that an equivalent study (performed in some other countries with domestic beer and that imported from the United States) would show similar conclusions, only in reverse. In more than one country in Europe, the Americana thing is sexy and youthful and energizing.

The amazing thing is that the imported beers quaffed in the aforementioned study in California displayed aged characteristics (cardboard, skunk) that are universally decried as being obnoxious by those in the brewing profession. And yet, because of the label, drinkers preferred them. Surely, this says it all. Isn't this the reason why a mediocre bottle of wine with the label of an upscale chateau can fetch obscenely high prices whereas a technically excellent product with a two-dollar price tag is decried as being plonk?

The fact is that there are those in the wine world who would introduce the same principles of striving for consistency and control, yet are sometimes pilloried for their efforts. Leo McCloskey at Enologix seeks to show that through analysis and computer-based records, it is possible to roadmap wine according to specific characteristics that can be controlled. It seems, though, that art will always triumph over science in the winemaking. Charm over control.

There are those in the wine world – sadly, folks that are too often persona non grata – who recognize this. Doyen among them is Fred Franzia, who famously acquired the Charles Shaw brand rights in 1995 for $25,000 at a time when California had a little more than 350,000 acres planted to wine grapes. Seven years later, a further 200,000 acres of grapes had been added and Franzia moved in on the surplus to launch his wine, colloquially known as Two Buck Chuck, and a wine that, as I write, runs to 5.5 million cases per annum. The story is legion of how in a blind test, one of this collection was selected as a finalist in a prestigious wine tasting. California winemakers don't like to talk of it, any more than the French vintners relished the famous tasting when the California wines swept the awards. Yet, which of us wanting to impress our dinner guests would bring out a wine that everyone knows cost us only a couple of bucks? Fred Franzia sells some 20 million cases of all of his brands per year and truly is the consumer's champion, being outspoken about ludicrous prices in restaurants. He says that no wine is worth more than $10 a bottle. Many winemakers in places like Napa are crippled with debts and purveying over-priced beverages not obviously better (to most folk's perception) than those available from Mr. Franzia.

Cost

So step back to compare your purchases by different yardsticks. And not the fantastic charge you will be stung for in a snooty restaurant, rather, the cost to take a supposed middle-of-the-road product off the supermarket shelf.

Let's say we have bought a 750-mL bottle of respectable Cabernet Sauvignon at a mid-price $10 dollars a bottle. On the basis that each bottle contains six servings, then each glass is costing us around $1.65.

Next, we buy a six-pack of premium ale. We are charged $6. That's $1 per serving.

Say the red wine is 15 percent alcohol and the beer is 5 percent alcohol. Then that's 9 cents per mL of alcohol for the wine and 6 cents per mL for the beer.

I think it was the legendary Robert Mondavi who said: "It takes a lot of good beer to make good wine." The sentiment is that wine-makers like their beer and drink a lot of it while they wait for the wine to ferment. Personally, I prefer the sentiment from the six-teenth century that:

> Wine is but a single broth
> Ale is meat, drink and cloth.

Yet, it is simply fact that beer is down-market. Recently, I was hosting a group of brewers from a well-known large company and in the restaurant, we each ordered one of their premium products. For once, glasses arrived – but in each there was an olive rolling around. I inquired, only to be told that it was a "poor man's cocktail."

It is high time that such ludicrous imagery and assumptions are driven out. Sure, there are beers for the routine drinking occasion, to wet the whistle and enjoy. But in the range of products available, there is a rich diversity, a drink for every conceivable drinking occasion and for everybody, male or female, of all legal drinking ages.

Enjoyment, partying, and fun or tradition and nostalgia; there is a beer for the occasion. Beer – liquid bread – is very much in keeping with a rich quality of life.

Wine and beer – both wonderful beverages, sublime outcomes of humankind's oldest agricultural endeavors. They have much to learn from one another.

FURTHER READING

Beer

I am sufficiently egotistic to cite my own *Beer: Tap Into the Art and Science of Brewing* (Oxford University Press, Second Edition, 2003) as the correct introductory read on beer. Neither can I ignore the only book that overtly addresses the issues pertaining to beer in the diet, by the same author: *Beer: Health and Nutrition* (Blackwell Publishing, 2004). For those seeking the history of beer, go to Ian S. Hornsey's *A History of Beer and Brewing* (Royal Society of Chemistry, 2003). For beer styles, there is a whole series published by the Brewers Association (go to <http://www.beertown.org/>). Readers eager for more detailed texts on the production processes involved in passing from barley to beer at point of sale should choose among D. E. Briggs, P. A. Brookes, R. Stevens, and C. A. Boulton, *Brewing: Science and Practice* (Woodhead, 2005); F. G. Priest and G. G. Stewart, eds., *Handbook of Brewing* (CRC Press, Taylor and Francis, 2006); and C. W. Bamforth, ed., *Brewing: New Technologies* (Woodhead, 2006).

Wine

Two good reads on the history of wine are by Patrick McGovern, *Ancient Wine: The Search for the Origins of Viticulture* (Princeton University Press, 2003) and by Rod Phillips, *A Short History of Wine* (Harper Collins, 2000). Scientific issues are addressed in *Wine: A Scientific Exploration*, edited by M. Sandler and R. Pinder (Taylor and Francis, 2003) and by R. S. Jackson in *Wine Science: Principles, Practice, Perception* (Academic Press, 2000). R. B. Boulton, V. L. Singleton, L. F. Bisson, and R. E. Kunkee address viticulture and wine making per se in *Principles and Practices of Winemaking* (Springer, 1996). Wine flavor is addressed by R. J. Clarke and J. Bakker in *Wine: Flavor Chemistry* (Blackwell, 2004). Head to your local store to find too many books on wines, styles, and their countries of origin. You might consider Karen MacNeil's *The Wine Bible* (Workman, 2001) as one of the more economical and less pretentious of the genre. For wine and health, go to Gene Ford's *The Science of Healthy Drinking* (Wine Appreciation Guild, 2003).

INDEX